T0325692

Multiscale Structural Topology Optimization

Series Editor
Piotr Breitkopf

Multiscale Structural Topology Optimization

Liang Xia

First published 2016 in Great Britain and the United States by ISTE Press Ltd and Elsevier Ltd

ISTE Press Ltd
27-37 St George's Road
London SW19 4EU
UK

www.iste.co.uk

Elsevier Ltd
The Boulevard, Langford Lane
Kidlington, Oxford, OX5 1GB
UK

www.elsevier.com

Notices

Knowledge and best practice in this field are constantly changing. As new research and experience broaden our understanding, changes in research methods, professional practices, or medical treatment may become necessary.

Practitioners and researchers must always rely on their own experience and knowledge in evaluating and using any information, methods, compounds, or experiments described herein. In using such information or methods they should be mindful of their own safety and the safety of others, including parties for whom they have a professional responsibility.

To the fullest extent of the law, neither the Publisher nor the authors, contributors, or editors, assume any liability for any injury and/or damage to persons or property as a matter of products liability, negligence or otherwise, or from any use or operation of any methods, products, instructions, or ideas contained in the material herein.

For information on all our publications visit our website at http://store.elsevier.com/

British Library Cataloguing-in-Publication Data
A CIP record for this book is available from the British Library
Library of Congress Cataloging in Publication Data
A catalog record for this book is available from the Library of Congress
ISBN 978-1-78548-100-0

Printed and bound in the UK and US

Contents

Introduction

In this chapter, the background and motivations of the book are first presented in section I.1. Literature reviews on related subjects, including linear and nonlinear topology optimization, multiscale modeling methods, model reduction strategies in multiscale modeling and simultaneous structure and materials design, are given in section I.2. Finally, the outline of the book is presented in section I.3.

I.1. Background and motivations

Various optimization methods for structural size, shape and topology designs have been developed and widely employed in engineering applications. Among these, topology optimization has been recognized as one of the most effective tools for least-weight and performance design, especially in aeronautics and aerospace engineering [ZHU 15]. Most existing research of topology optimization focuses on the design of monoscale structures; in other words, the considered structures are made of homogeneous materials.

In the recent years, there is an increasing use of high-performance heterogeneous materials such as fibrous composite, concrete, metallic porous material and metal alloy for their advantageous overall characteristics, which result in superior structural mechanical responses and service performance. Although from the structural level point of view these materials can be considered homogeneous and conventional design approaches for homogeneous structures can still be used, the pronounced heterogeneities have significant impacts on the structural behavior. Therefore, in order to allow for reliable mechanical designs, we need to account for material microscopic heterogeneities and constituent behaviors so as to accurately assess structural performance.

Meanwhile, the fast progress made in the field of materials science allows us to control the material microstructure composition to an unprecedented extent [FUL 10]. The overall behavior of heterogeneous materials depends strongly on the size, shape, spatial distribution and properties of the constituents. With all these in mind, we come up naturally with the idea that designing materials simultaneously along with the design of structures would result in higher performance structures. In addition, the recently emerging and rapidly developing techniques of three-dimensional (3D) printing or additive manufacturing, such as fused deposition modeling, stereolithography and selective laser sintering, provide the capability of manufacturing extremely fine and complex microstructures, which make it possible to generate more innovative, lightweight and structurally efficient designs.

I.1.1. *Motivation 1*

The primary motivation of the book is to make a first attempt toward topology optimization design of nonlinear highly heterogeneous structures. Generally speaking, topology optimization design of multiscale structures (Figure I.1) can be viewed as an extension of conventional monoscale design except that the material constitutive law is governed by one or multiple representative volume elements (RVEs) defined at the microscopic scale. In the case of linear elasticity, topology optimization design of a structure made of the RVE is a rather straightforward application of conventional linear design routine [SIG 01, HUA 10a], because the effective or homogenized constitutive behavior of the considered RVEs can be explicitly determined by homogenization analysis.

When nonlinearities are present at the microscopic scale, i.e. nonlinear RVEs under consideration, topology optimization design of such multiscale structures is a rather challenging task. First, as will be discussed subsequently in section I.2.1, nonlinear topology optimization is not at all a trivial task even for homogeneous structures due to the increased computing cost and the required solution stabilization schemes, not to mention highly heterogeneous multiscale structures. Second, the multiscale dilemma in terms of heavy computational burden (see section I.2.2) is even more pronounced in topology optimization: not only is it required to solve the time-consuming multiscale problem once, but for many different realizations of the structural topology. For these reasons, there has been very limited research in the literature on topology optimization design of multiscale nonlinear structures before the recent works from the author and his collaborators [XIA 14b, XIA 15c, FRI 15b].

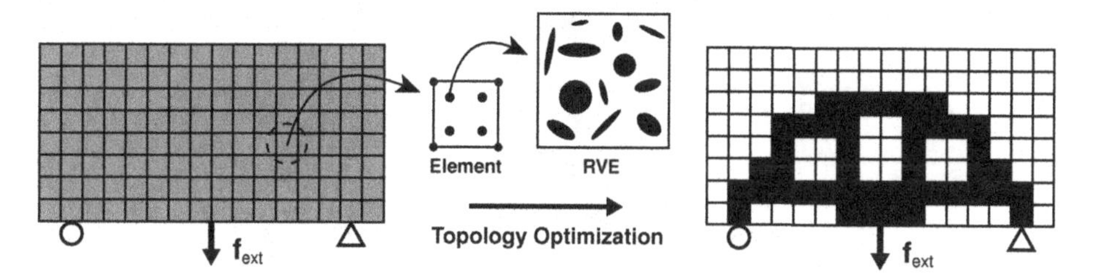

Figure I.1. *Illustration of topology optimization of multiscale structures [XIA 14b]*

I.1.2. *Motivation 2*

The second motivation of the book is to design the topologies of both macroscopic structure and microscopic materials simultaneously. In other words, by topology optimization one determines not only the optimal spatial material layout distribution at the macroscopic structural scale, but also the optimal local use of the cellular material at the microscopic scale, as schematically shown in Figure I.2. The subject of simultaneous topology optimization design of both structure and materials can be found in earlier works [THE 99, ROD 02, ZHA 06]. In this book, we revisit the subject with an emphasis on solving the nonlinear scale interface equilibrium within a multiscale analysis framework [XIA 14a, XIA 15b].

Concerning simultaneous design, the microscale material topologies are optimized in response to the macroscale equilibrium solution; the optimized materials in turn result in a variation of the macroscale constitutive behavior. The equilibrium of the scale interface is therefore nonlinear, which has been well acknowledged [THE 99, BEN 03]. However, in practice it has never been specifically dealt with in previous works [XIA 14a, XIA 15b]; in particular, the microscale material design is treated integrally as a generalized nonlinear constitutive behavior and the nonlinear equilibrium due to the locally optimized or adapted materials is solved within a nonlinear multiscale analysis framework. One particular advantage of doing so is that we can improve the design efficiency by straightforward application of the existing model reduction strategies for nonlinear heterogeneous materials [XIA 15b].

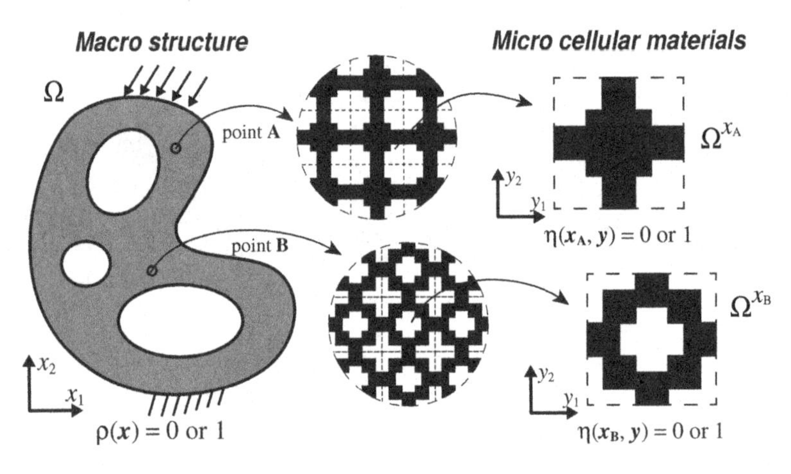

Figure I.2. *Illustration of simultaneous topology optimization of structure and materials [XIA 14a]*

I.1.3. *Motivation 3*

The third motivation of the book is to develop computer codes for the designs of nonlinear structures and materials with extreme constitutive properties. The compactly developed codes including all ingredients for topology optimization would be beneficial in terms of presenting the methods and attracting more research interests. In addition, the compactness and extensibility of the codes make it possible to serve for further academic researches and also for educational purposes.

This motivation is inspired by the 99-line Matlab code in the seminal article by Sigmund [SIG 01] and other subsequent educational codes [HUA 10a, CHA 10, AND 11, TAL 12a], which have significantly contributed to the popularity and to the development of topology optimization. We have also benefited from these educational papers. For instance, in Chapter 2 [XIA 14b] we apply the discrete level set method [CHA 10] to carry out topology optimization. Chapters 1 and 3 on the design of multiscale structures [XIA 15c, FRI 15b] and Chapters 4 and 5 on the simultaneous design of structure and materials [XIA 14a, XIA 15b] are all built on top of the 88-line code framework [AND 11] along with the evolutionary design scheme [HUA 10a]. In the Appendix, we provide one of the developed codes for the design of extreme materials in Matlab.

I.2. Literature review on related subjects

In the following, section I.2.1 gives a literature review on the developments of linear and nonlinear topology optimization. Section I.2.2 presents multiscale modeling methods and the reduced-order modeling (ROM) strategies for them. Section I.2.3 reviews conventional approaches on simultaneous topology optimization of structure and materials together with our recently proposed alternative approach.

I.2.1. *Topology optimization*

Since the seminal paper by Bendsøe and Kikuchi [BEN 88], topology optimization has undergone a remarkable development in both academic research [BEN 03, HUA 10b, DEA 14] and industrial applications [ZHU 15]. Various approaches have been proposed, such as density-based methods [BEN 89, ZHO 91], evolutionary procedures [XIE 93, XIE 97, ZHU 07, HUA 09] and level set methods [SET 00, WAN 03, ALL 04b, BUR 04], all with the same purpose of finding an optimal structural topology or material layout within a given design domain subjected to constraints and boundary conditions as is shown in Figure I.3. The key merit of this approach over shape or sizing optimizations is that it does not require a presumed topology.

Originally, topology optimization was considered as a 0-1 discrete problem or a binary design setting, which is known as ill conditioned upon Kohn and Strang [KOH 86]. The major challenge lies in solving a large-scale integer programing problem, where the high computing cost typically precludes the use of gradient-free algorithms. Bendsøe and Kikuchi [BEN 88] relaxed the problem by assuming porous microstructures at a lower separated scale. Similar treatments were followed in [GUE 90, SUZ 91, ALL 04a].

Shortly after, Bendsøe proposed another density-based method with a much more simplified assumption [BEN 89], also known as solid isotropic material with penalty (SIMP) [ZHO 91, ROZ 01].

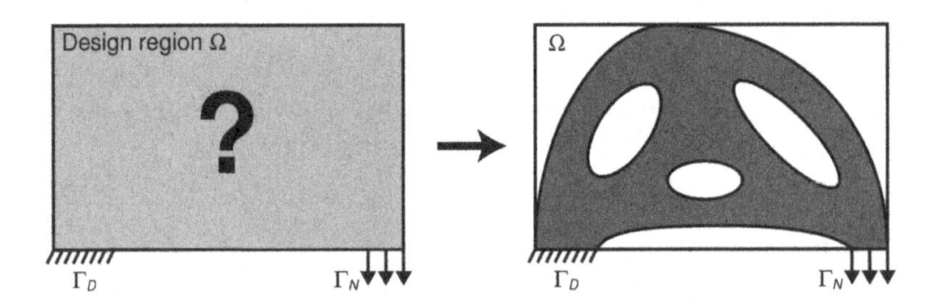

Figure I.3. *Illustration of structural topology optimization [XIA 15c]*

Another popular approach, named evolutionary structural optimization (ESO), was initially proposed by Xie and Steven [XIE 93] based on a heuristic element removal strategy. It has been reported that the ESO method corresponds to a sequential linear programing approximate method for certain cases [TAN 02]. Despite its popularity [XIE 97], the ESO method was also largely argued [SIG 01] or even criticized [ZHO 01] mainly for lacking restitution mechanisms for material removal. To circumvent the deficiencies, bidirectional evolutionary structural optimization (BESO) [QUE 00, ZHU 07, HUA 09] and soft-killing BESO [HUA 10a, HUA 10b] were later developed, allowing both material removal and addition.

Apart from the abovementioned approaches, there exist several alternative approaches such as the bubble method [ESC 94], topological derivative [SOK 99], level-set method, [SET 00, WAN 03, ALL 04b, BUR 04, VAN 13] and phase filed method [BOU 03]. For a more critical review and detailed comparison of the various design approaches, the reader may refer to [SIG 13].

Early works on topology optimization were restricted to linear structural designs. Continuous efforts have been conducted to extend topology optimization to nonlinear structural designs considering various sources of nonlinearity, such as geometrical nonlinearity [BUH 00, GEA 01, PED 01, BRU 03, YOO 05, WAN 14a, LUO 15], material nonlinearity [YUG 95, BEN 96, MAU 98, YUG 99, SCH 01, YOO 07, BOG 12] and both geometrical and material nonlinearities simultaneously [JUN 04, HUA 07b, HUA 08]. In the case of geometrical nonlinearity, the main difficulty of nonlinear topology optimization arises primarily from the existence of compliant or ersatz materials at void regions. The excessively distorted elements at void regions during topology optimization result in the instability of Newton–Raphson (NR) solution process [BUH 00, PED 01, BRU 03, YOO 05]. To stabilize the numerical solution, several remedy schemes have been proposed by, for instance, excluding internal forces of void elements [BUH 00], removing low-density elements [BRU 03], including additional connectivity parameterization [YOO 05] or, recently, a variant interpolation scheme [WAN 14a]. In the case of material nonlinearity, such as plasticity, we need to define material interpolations between the intermediate density values and elastic material modulus, plastic hardening modulus and yield stress when employing density-based methods [MAU 98, YUG 99, SCH 01, YOO 07, BOG 12, KAT 15]. These material interpolation models need to be adjusted carefully so as to guarantee optimization solutions, while the choice of them lacks physical interpretation. As compared to continuously defined methods, ESO-type methods [HUA 10b] and discrete level-set methods [CHA 10, WEI 10] naturally omit the definition of supplementing pseudo-relationships between intermediate densities and their constitutive behaviors for the sake of their discrete nature, resulting in algorithmic advantages. The robustness and performance of discrete methods have been shown for the design of nonlinear structures [HUA 07b, HUA 08] and recently for nonlinear multiscale structures [XIA 14b, XIA 15c, FRI 15b].

I.2.2. *Multiscale modeling*

Brute force approaches such as directly modeling the microstructure at the coarse scale model are practically not feasible because of the prohibitive computational expense. Instead, homogenization is usually applied to bridge both structural and material scales [GUE 90, HAS 98a, HAS 98b, MIC 99]. The key hypotheses of homogenization are the separation of scales and the periodicity, as is shown in Figure I.4. It is assumed that the microscopic length scale is much smaller than the macroscopic length scale such that the RVE model can be considered as periodically ordered pattern, while the RVE is large enough to be dealt with using the continuum mechanics theory. By means of homogenization, we may evaluate the effective or homogenized constitutive behavior of the considered RVE and then use it to serve marcroscale assessment [GUE 08, XU 11, XU 12, XIA 13a].

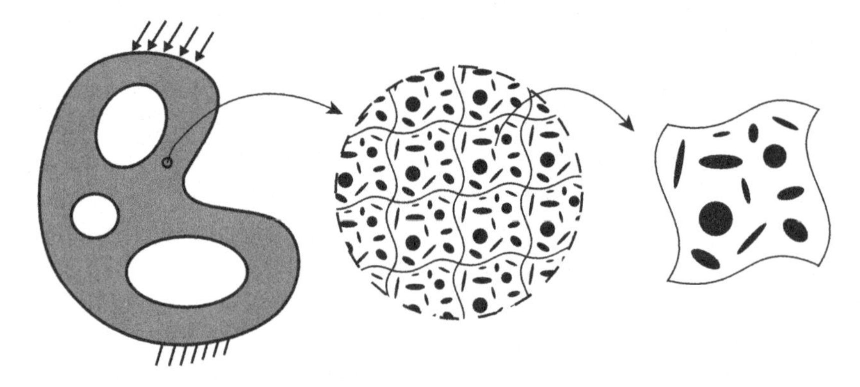

Figure I.4. *Illustration of a two-scale structure and periodically patterned RVE [XIA 14b]*

However, this approach encounters difficulties when geometrical and physical nonlinearities are present at the material scale. For this reason, computational homogenization approaches have been proposed [SMI 98, MIE 99, FEY 00, KOU 01, GHO 01, MIE 02, IBR 03, IBR 10] and largely developed in the last decade [GEE 10] in order to assess the macroscopic influence of microscopic heterogeneities and nonlinearities. Note that within the finite element analysis framework, this approach is also known as the FE square (FE^2) method [FEY 00]. In general, it asserts that each point of the macroscopic discretization is associated with a RVE of the (nonlinear) microstructured material. Then for each macroscopic equilibrium iteration, a nonlinear load increment needs to be computed for each of the (many) RVEs. In return, the average stress across the RVE is then used as the macroscopic stress tensor without requiring explicit constitutive relations at the macroscopic scale.

A downside of this very general FE^2 method is the high computational burden. First, many nonlinear load steps need to be computed at the microscopic level that leads to a prohibitive amount of computing time. Second, when path-dependent constitutive behaviors are considered at the microscopic scale, the microscopic degrees of freedom and the history variables describing the material state need to be stored for each point within each RVE, which leads to a significant amount of additional storage requirements. One straightforward solution to alleviate the computational requirements is parallel computing [FEY 00, MOS 14], as microscale problems are independent and embarrassingly parallel. Note that the implementation of parallel computing contributes significantly in terms of limiting the computing time, but does not necessarily reduce the computing cost due to additional interchanges between the two scales. For this reason, we need to turn to alternative strategies by means of model reduction or simplification.

ROM has been systematically researched and widely used in the fields of computational mechanics in order to reduce computing cost, data storage requirements and computing time [QUE 05, FOR 09]. Some other applications can also be found for the representation of material mircrostructure [GAN 07, XIA 13a] and structural optimization design [XIA 10, RAG 13]. In terms of reducing the computing effort for the evaluation of nonlinear RVEs at the microscopic scale, numerous ROMs can be found in the literature for the representation or approximation of the effective constitutive behavior of nonlinear heterogeneous materials, using reduction strategies such as proper orthogonal decomposition (POD) [YVO 07], proper generalized decomposition [LAM 10, ELH 13, CRE 13], hyper reduction [MIL 13, HER 14], material map model [TEM 07b], eigen deformation-based reduction [OSK 07, YUA 09], nonuniform transformation field analysis (NTFA) [MIC 03, MIC 04] and numerical explicit potentials (NEXP) [YVO 09, YVO 13, LE 15]. Note that, by simultaneous use of parallel computing and ROM [FRI 13b], a further reduction in computational time can be achieved in multiscale analysis [FRI 14].

In the case of elasticity, the database-type methods, such as the material map model [TEM 07b, TEM 07a] and NEXP [YVO 09], have shown promising performances in terms of both modeling accuracy and computing efficiency. The general idea of this type of methods is to compute off-line a certain number of RVE problems as a database, then the effective RVE behavior is approximated using the precomputed database by means of interpolation schemes. The on-line macroscale computation then directly uses the cheaper approximated constitutive behavior of the RVE without demanding to solve full-scale RVE problems. These methods apply for viscoelastic materials [TRA 11] and nonlinear hyperelastic materials

[TEM 07b, TEM 07a, YVO 13], and have also been extended for stochastic nonlinear elastic materials [CLÉ 12, CLÉ 13]. The development of ROMs for the representation of RVE involving path-dependent constitutive laws, such as plasticity, is a more challenging subject under development. Here, we refer to the NTFA method [MIC 03, MIC 04, FRI 11, FRI 13a] and adaptive POD approach as implemented in [MIL 13, HER 14]. In particular, NTFA reduces the inelastic strain field found in a microstructured material to a finite dimensional but spatially heterogeneous basis of nonuniform transformation strains. Generally speaking, all these established ROMs apply straightforwardly to topology optimization of multiscale nonlinear structures as long as the macroscopic equilibrium solution is provided.

I.2.3. *Simultaneous structure and materials design*

Topology optimization has not only been applied for structural designs, but also for material microstructural design [CAD 13]. By means of inverse homogenization, the SIMP method has also been used for designing material microstructures with prescribed constitutive properties [SIG 94, SIG 00], extreme thermal expansion coefficients [SIG 97, GIB 00], extreme viscoelastic behavior [YI 00, AND 14a, CHE 14, HUA 15], maximum stiffness and fluid permeability [GUE 06, GUE 07] and, recently, hyperelastic properties [WAN 14b]. Similar problems have also been addressed by level-set methods [CHA 08, CHA 12] and ESO-type methods [HUA 11, HUA 12]. Some other works [NEV 00, FUJ 01, NEV 02, ZHA 07, NIU 09, SU 10, GU 12, KAT 14, AND 14b] also fall within this context. Up until now, topology optimization design of materials with extreme constitutive properties (Figure I.5) has followed a rather standard routine [XIA 15a].

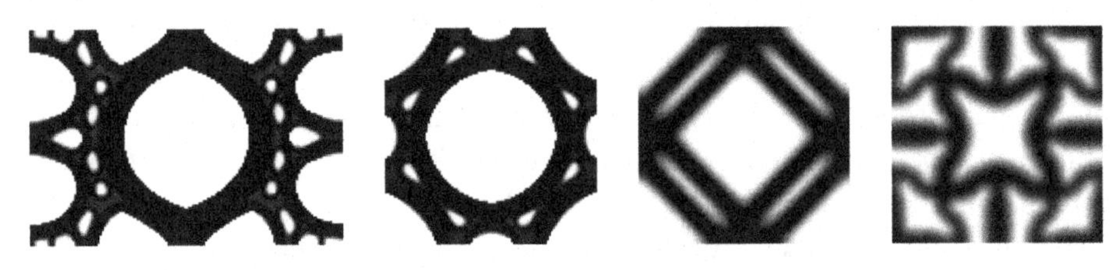

Figure I.5. *Material microstructures (50% porosity): the first two with extreme bulk moduli, the third with extreme shear modulus and the last one with negative Poisson's ratio [XIA 15a]*

With the established material microstructural design models, we naturally came up with the idea of simultaneous or integrated designs of both material and structure. The most commonly applied strategy is designing a universal material microstructure at the microscopic scale either for a fixed [SU 10, HUA 13] or simultaneously changed [DEN 13, ZUO 13, YAN 14, XU 15, GUO 15] structure at the macroscopic scale. Zhang *et al.* [ZHA 06] went a step further by designing several different cellular materials for different layers of a layered structure. In fact, an earlier attempt can be traced back to [ROD 02], where simultaneous optimal designs were performed for both structure and element wisely varying cellular materials following a decomposed design procedure [BEN 95, THE 99]. This work has later been extended to 3D [COE 08] and to account for hyperelasticity [NAK 13]. Some more specific simultaneous designs can be found in [SET 05, SET 06, LIU 12, GAO 12, GAO 13, COE 15] for structural topology and composite laminate orientations and in [LV 14] for structural topology and closed liquid cell materials. Another recent work, in [ALE 15], demonstrates the efficient solution of the design problem by using a spectral coarse basis preconditioner and without assuming the separation of length scales.

Due to the intensive computational cost, the nonlinear scale-interface equilibrium [THE 99, BEN 03] of the locally optimized or adapted materials was neglected in early works [ROD 02, COE 08, ZHA 06], conjecturing that both scale design variables were updated simultaneously and no converged local material design results were required for macroscopic structural equilibrium. Unlike previous design approaches, this nonlinearity has not been neglected but is specially addressed in our recent work [XIA 14a] treating the microscale material design integrally as a generalized nonlinear constitutive behavior, and in which the nonlinear scale-interface equilibrium problem is resolved by the FE^2 method. In precise, complete microscale optimizations are solved for each macroscale displacement increment. The nonlinear scale-interface equilibrium is searched by means of the NR method using the effective stresses evaluated on the optimized material microstructural topologies. It has been shown that this FE^2-based design approach can provide similar topology solutions in comparison to the iterative design approach [HUA 13, YAN 14], while requiring much less computing cost due to the reduced interchange between the two scales. Another advantage of treating the material optimization process as a generalized constitutive behavior is that the existing model reduction strategies for nonlinear heterogeneous materials can be applied straightforwardly to improve design efficiency, as we have shown in [XIA 15b].

I.3. Outline of the book

This book is organized as follows:

In this chapter, we have presented our motivations and the attempts we will make to move toward topology optimization design of multiscale nonlinear structures, including design of highly heterogeneous structures, material microstructural design and simultaneous design of structure and materials. We have also reviewed the state of the art of the related subjects, which appears in [XIA 15c].

In Chapter 1, we primarily develop a multiscale design framework for topology optimization of multiscale nonlinear structures. Highly heterogeneous structures made up of nonlinear elastic RVE are considered. Conventional first-order computational homogenization FE^2 method is adopted to bridge the two separate scales. Topology optimization is carried out using the BESO method for macroscopic structural stiffness maximization with a constraint on material volume usage. For the sake of its discrete nature, the BESO method naturally omits the definition of supplementing pseudo-relationships between intermediate densities and the moduli as required when employing SIMP-type models, resulting in algorithmic advantages, especially in dealing with multiscale nonlinear structures. In contrast to the conventional nonlinear design of homogeneous structures, this design framework provides an automatic design tool by which the considered material models can be governed directly by the realistic microstructural geometry and the microscopic constitutive models. The content of this chapter comes primarily from [XIA 14b] and [XIA 15c].

Intending to alleviate the heavy computational burden of the design framework presented in Chapter 1, we develop a POD-based adaptive surrogate model in Chapter 2 for the RVE solutions at the microscopic scale. We note that the optimization process requires multiple design loops involving similar or even repeated computations at the microscopic scale, which perfectly suits the surrogate learning process. The surrogate model is constructed in an on-line manner: initially built by the first design iteration is then updated adaptively in the subsequent design iterations. The displacement fields are treated as snapshots and a reduced basis is obtained by snapshot POD. The surrogate model is built by interpolating the POD projection coefficients in terms of the effective strain using diffuse approximation. The surrogate model has shown promising performance in terms of reducing computing cost and modeling accuracy when applied to the design framework for nonlinear elastic cases. The content of this chapter has been given in [XIA 14b].

In Chapter 3, we take a step further toward the design of multiscale elastoviscoplastic structures. In order to realize the design in realistic computing time and with affordable memory requirement, we directly employ an established method, the potential-based reduced basis model order reduction (pRBMOR) with graphics processing unit (GPU) acceleration [FRI 14], to resolve the microscopic problems for this severe material nonlinearity. The pRBMOR implementation allows for memory and CPU time savings by factors of 10^5 and beyond with respect to FE simulations. With regard to topology optimization, the sensitivities of the design variables for nonlinear dissipative problems are derived in a clear and rigorous manner using the adjoint method. In addition, a stabilization scheme controlling the number of recovered elements is implemented to enhance previous versions of BESO updating scheme for linear designs. The content of this chapter has been given in [FRI 15b].

So far we have always assumed a fixed material microstructure at the microscopic scale. In Chapter 4, we are focused on the simultaneous design of both macroscopic structure and microscopic materials. Within the same established multiscale design framework, we define topology variables and volume fraction constraints at both scales. In this model, the material microstructures are optimized in response to the macroscopic solution, which results in the nonlinearity of the equilibrium problem of the interface of the two scales. This model treats the optimization process of material microstructure as a generalized nonlinear constitutive behavior, and the nonlinear equilibrium problem can be resolved naturally within the multiscale design framework by the FE^2 method. The proposed model allows us to obtain optimal structures with spatially varying properties realized by the simultaneous design of microstructures at the microscopic scale. Note that the designed structure with varying constitutive behaviors due to the microstructures are in fact constituted by only one base material, which greatly favors 3D printing setting that a single material can usually be used for fabrication. The content of this chapter has been given in [XIA 14a].

Treating the material optimization process as a generalized constitutive behavior (Chapter 4) allows us to improve the design efficiency drastically by straightforward application of the existing ROMs for nonlinear materials. In Chapter 5 we apply a reduced database model, NEXP [YVO 09], to approximate this generalized constitutive behavior. By this model, we build *a priori* a database from a set of numerical experiments in the space of effective strain. Each value in the database corresponds to the strain energy density evaluated on a material microstructure, optimized in accordance to the imposed effective strain value. By tensor decomposition, a continuous representation of the strain energy density is built as a sum of products of one-dimensional interpolation functions. As a result of this off-line phase, we have the effective strain–energy and stress–strain relations required for the macroscopic equilibrium solution given in the explicit manner. The explicit NEXP approximation can serve the simultaneous design purpose providing ultraresolution structures at a significantly reduced computational cost. The content of this chapter has been given in [XIA 15b].

In section 5.7, we conclude the book and give perspectives on future research.

The developments in the book have given rise to a number of prototype computer codes. We include in the Appendix our recent educational article [XIA 15a] regarding design of materials using topology optimization and energy-based homogenization approach together with a Matlab code.

The chapters are presented in order of progressive difficulty: from linear to nonlinear problems, from uniform to spatially varying heterogeneity with progressive introduction of reduced-order modeling strategies. However, each chapter is given independency, where physical definitions and assumptions are reinvoked when necessary at the beginning of successive chapters.

1

Topology Optimization Framework for Multiscale Nonlinear Structures

This chapter introduces a topology optimization framework for multiscale nonlinear structures. As an extension of the conventional monoscale design, except that the material constitutive law is governed by one or multiple RVEs defined at the microscopic scale, the developed general multiscale design framework is made up of two key ingredients: *multiscale modeling* for structural performance simulation and *topology optimization* for structural design. This framework will serve as a basis for the developments presented in the subsequent chapters.

With regard to the first ingredient, we employ the first-order computational homogenization method FE^2 [FEY 00] to bridge structural and material scales. By this method, a clear separation of scales is asserted and periodicity is assumed at the microscopic scale. With regard to the second ingredient, we apply the BESO method for its robustness and the performance of the resulting structures in nonlinear designs [HUA 08, HUA 10b]. The BESO method is based on an evolutionary mechanism that the topological change is realized by gradual material removal and addition.

In the following, in section 1.1 we first summarize and review the FE^2 method. The implementation of a unified periodic boundary condition (PBC) [XIA 03] is given in section 1.2. Finite element discretization formulations are presented in section 1.3. The FE^2-based nonlinear topology optimization model using the BESO method is given in section 1.4. The summarized general multiscale design framework is presented in section 1.5. Section 1.6 carries out the designs of a two-scale cantilever structure made of periodically patterned anisotropic short-fiber reinforced composite with nonlinear elastic behaviors and the nonlinear design results are compared with monoscale design results of the corresponding homogeneous structures. We give concluding remarks in section 1.7.

1.1. FE^2 method

The FE^2 method assumes the hypothesis of the separation of scales and periodicity as already shown in Figure I.4. By finite element analysis (FEA), each material point (Gauss integration point) at the macroscopic scale is attributed with a prescribed RVE. At the macroscopic scale, the material appears to be homogeneous but with unknown mechanical properties. These mechanical properties are related to the heterogeneities of the RVE at the microscopic scale, which contribute strongly to the overall mechanical response observed at the larger scale.

Let $\boldsymbol{x}$ and $\boldsymbol{y}$ denote the position of a point at the macro- and microscales, respectively. Within the macroscopic domain Ω, the macroscopic displacement $\bar{\boldsymbol{u}}(\boldsymbol{x})$, the macroscopic strain $\bar{\boldsymbol{\varepsilon}}(\boldsymbol{x})$ and the macroscopic stress $\bar{\boldsymbol{\sigma}}(\boldsymbol{x})$ are considered. Their microscopic counterparts at the microscopic scale are the displacement $\boldsymbol{u}(\boldsymbol{x},\boldsymbol{y})$, the infinitesimal strain $\boldsymbol{\varepsilon}(\boldsymbol{x},\boldsymbol{y})$ and the stress $\boldsymbol{\sigma}(\boldsymbol{x},\boldsymbol{y})$. While the constitutive model for each material phase of the RVE at the microscopic scale is assumed to be known, the explicit macroscale constitutive relations that can account for the microstructural heterogeneities are rarely ever at hand. Therefore, the macroscale stress can often only be computed as a function of the microscale stress state by means of volume averaging over the associated RVE domain Ω_x through

$$\bar{\boldsymbol{\sigma}}(\boldsymbol{x}) = \langle \boldsymbol{\sigma}(\boldsymbol{x},\boldsymbol{y}) \rangle = \frac{1}{|\Omega_x|} \int_{\Omega_x} \boldsymbol{\sigma}(\boldsymbol{x},\boldsymbol{y}) \, \mathrm{d}\Omega_x, \qquad [1.1]$$

in which $\boldsymbol{\sigma}(\boldsymbol{x},\boldsymbol{y})$ is evaluated by solving the boundary value problem of the RVE by constraining $\langle \boldsymbol{\varepsilon}(\boldsymbol{x},\boldsymbol{y}) \rangle$ equal to $\bar{\boldsymbol{\varepsilon}}(\boldsymbol{x})$, i.e.,

$$\bar{\boldsymbol{\varepsilon}}(\boldsymbol{x}) = \langle \boldsymbol{\varepsilon}(\boldsymbol{x},\boldsymbol{y}) \rangle = \frac{1}{|\Omega_x|} \int_{\Omega_x} \boldsymbol{\varepsilon}(\boldsymbol{x},\boldsymbol{y}) \, \mathrm{d}\Omega_x, \qquad [1.2]$$

where PBC are usually applied to define this constraint in accordance with the assumed periodicity assumption. Note that when cracks, voids and rigid inhomogeneities are present in the RVE, the foregoing definitions for the macroscopic stress and stain tensors need to be extended [MIC 99].

In addition to the scale bridging relations [1.1] and [1.2], the macroscopic stress $\bar{\boldsymbol{\sigma}}$ needs to be in equilibrium with the external tractions $\bar{\boldsymbol{t}}_*$ on the Neumann boundary $\partial\Omega_{\mathrm{N}}$ (body forces are out of consideration) and the displacements have to satisfy the Dirichlet conditions $\bar{\boldsymbol{u}} = \bar{\boldsymbol{u}}_*$ on $\partial\Omega_{\mathrm{D}}$:

$$\operatorname{div} \bar{\boldsymbol{\sigma}}(\boldsymbol{x}) = \mathbf{0} \ \text{in}\ \Omega, \quad \bar{\boldsymbol{\sigma}} \cdot \bar{\boldsymbol{n}} = \bar{\boldsymbol{t}} = \bar{\boldsymbol{t}}_* \ \text{on}\ \partial\Omega_{\mathrm{N}}, \quad \bar{\boldsymbol{u}} = \bar{\boldsymbol{u}}_* \ \text{on}\ \partial\Omega_{\mathrm{D}}. \qquad [1.3]$$

A schematic illustration of the FE^2 method is given in Figure 1.1 and can also be seen later in Figure 1.3. In summary, the FE^2 method consists of the following steps:

1) evaluate the macroscale strain $\bar{\boldsymbol{\varepsilon}}(\boldsymbol{x})$ with an initially defined setting;

2) define PBC on the associated RVE according to $\bar{\boldsymbol{\varepsilon}}(\boldsymbol{x})$;

3) evaluate the microscale stress $\boldsymbol{\sigma}(\boldsymbol{x}, \boldsymbol{y})$ by solving the RVE problem;

4) compute the macroscale stress $\bar{\boldsymbol{\sigma}}(\boldsymbol{x})$ via volume averaging $\boldsymbol{\sigma}(\boldsymbol{x}, \boldsymbol{y})$;

5) evaluate the tangent stiffness tensor $\bar{\mathbb{C}}^{\tan}(\boldsymbol{x})$ at the macroscale point $\boldsymbol{x}$;

6) update the displacement $\bar{\boldsymbol{u}}(\boldsymbol{x})$ using the NR method;

7) repeat steps 2–6 until the macroscale equilibrium is achieved.

Note that it is usual that there exists no explicit closed-form expression of $\bar{\mathbb{C}}^{\tan}$ at the macroscopic scale when nonlinearities are present at the microscopic scale. One possible but time-consuming solution is to approximate it using a perturbation method [FEY 00]. To be precise, the structural response is evaluated for a small strain variation $\boldsymbol{\delta\varepsilon}$ at the converged solution. The tangent stiffness tensor $\bar{\mathbb{C}}^{\tan}$ could be reconstructed by exciting each of the components of $\boldsymbol{\delta\varepsilon}$. An alternative perturbation approach based on nodal condensation is given in [KOU 01]. It is worth noting that computing the tangent stiffness tensor by the perturbation method requires solving the RVE problem four (in 2D) or six (in 3D) additional times, the cost of which is not negligible. For this reason, it is also suggested to use the initial elastic stiffness tensor during the NR solution process [FEY 00, YVO 07].

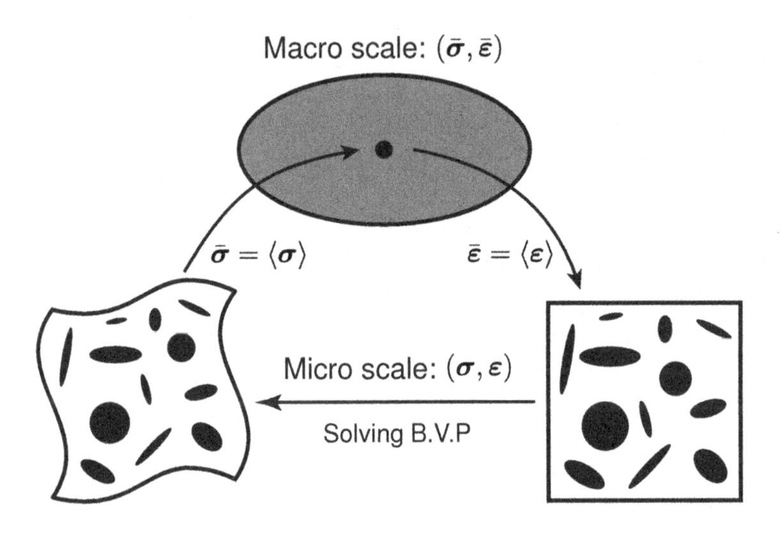

Figure 1.1. *Illustration of first-order computational homogenization scheme [XIA 15c]*

1.2. Periodic boundary conditions

The formulation of the periodic RVE can be derived in a systematic way using the two-scale asymptotic expansion method [GUE 90], or following a process which is also valid for random media [MIC 99], which is followed in this work. Upon the assumption of periodicity at the microscopic scale, the displacement field of the microstructure can be written as the sum of a macroscopic displacement field and a periodic fluctuation field $\boldsymbol{u}^*$ [MIC 99]

$$\boldsymbol{u}(\boldsymbol{x}, \boldsymbol{y}) = \bar{\boldsymbol{\varepsilon}}(\boldsymbol{x}) \cdot \boldsymbol{y} + \boldsymbol{u}^*(\boldsymbol{y}). \quad [1.4]$$

Because $\boldsymbol{u}^*(\boldsymbol{y})$ is periodic, its strain average $\langle \boldsymbol{\varepsilon}(\boldsymbol{u}^*) \rangle$ equals zero, and the average of the microscale strain thus equals directly to the macroscale strain

$$\langle \boldsymbol{\varepsilon}(\boldsymbol{x}, \boldsymbol{y}) \rangle = \bar{\boldsymbol{\varepsilon}}(\boldsymbol{x}). \quad [1.5]$$

In order to compute the microscopic stress field, the boundary value problem induced by an overall strain $\bar{\varepsilon}(\boldsymbol{x})$ has to be solved at the microscopic scale on the RVE: find $\boldsymbol{\sigma}, \varepsilon, \boldsymbol{u}^*$ such that

$$\begin{cases} \boldsymbol{\sigma}(\boldsymbol{x}, \boldsymbol{y}) = \mathbb{C}(\boldsymbol{y}) : (\bar{\varepsilon}(\boldsymbol{x}) + \varepsilon(\boldsymbol{u}^*(\boldsymbol{y}))) \\ \text{div } \boldsymbol{\sigma}(\boldsymbol{x}, \boldsymbol{y}) = \boldsymbol{0}, \ \boldsymbol{u}^* \text{ periodic}, \ \boldsymbol{\sigma} \cdot \boldsymbol{n} \text{ anti-periodic}, \end{cases} \quad [1.6]$$

where $\mathbb{C}(\boldsymbol{y})$ is the linear elastic tensor at the microscopic scale; "periodic" indicates that all components of $\boldsymbol{u}^*$ take identical values on points of the opposite sides of the boundary $\partial\Omega_x$; "anti-periodic" indicates that all components of $\boldsymbol{\sigma} \cdot \boldsymbol{n}$ take opposite values on points of the opposite sides of $\partial\Omega_x$, deduced from the periodicity of $\boldsymbol{\sigma}$ and the fact that the normal vectors $\boldsymbol{n}$ at opposite sides of $\partial\Omega_x$ are opposite.

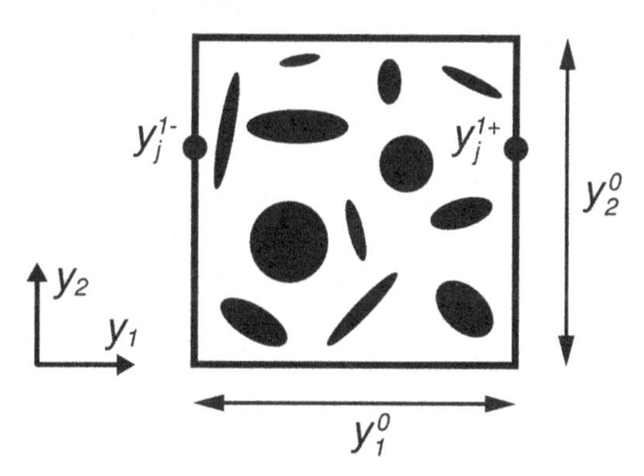

Figure 1.2. *An illustrative 2D rectangular RVE [XIA 14b]*

In practice, equation [1.4] cannot be directly applied to the boundaries since the periodic fluctuation part $\boldsymbol{u}^*$ is generally unknown. The general expression is usually transformed into a certain number of explicit constraints between the corresponding pairs of nodes on the opposite surfaces of the RVE [XIA 03]. Consider an RVE such as one periodic cellular microstructure in Figure 1.2, the displacements on a pair of opposite boundaries are

$$\begin{cases} \boldsymbol{u}(\boldsymbol{x}, \boldsymbol{y})^{k+} = \bar{\varepsilon}(\boldsymbol{x}) \cdot \boldsymbol{y}^{k+} + \boldsymbol{u}^*(\boldsymbol{y}) \\ \boldsymbol{u}(\boldsymbol{x}, \boldsymbol{y})^{k-} = \bar{\varepsilon}(\boldsymbol{x}) \cdot \boldsymbol{y}^{k-} + \boldsymbol{u}^*(\boldsymbol{y}), \end{cases} \quad [1.7]$$

where superscripts "$k+$" and "$k-$" denote the pair of two opposite parallel RVE boundary surfaces that are oriented perpendicular to the kth direction. The periodic fluctuation part $\boldsymbol{u}^*$ can be eliminated through the difference between the displacements on the two opposite surfaces

$$\boldsymbol{u}^{k+} - \boldsymbol{u}^{k-} = \bar{\boldsymbol{\varepsilon}}(\boldsymbol{x}) \cdot (\boldsymbol{y}^{k+} - \boldsymbol{y}^{k-}), \qquad [1.8]$$

in which the difference $\Delta \boldsymbol{y}_k = (\boldsymbol{y}^{k+} - \boldsymbol{y}^{k-})$ is constant for every pair of parallel boundary surfaces. With specified $\bar{\boldsymbol{\varepsilon}}(\boldsymbol{x})$, the right side becomes constant and such equations can be easily applied in the FEA as nodal displacement constraint equations.

The application of [1.8] automatically guarantees the "periodic" condition of $\boldsymbol{u}^*$ in [1.6]. As for the "anti-periodic" condition of the tractions along the boundaries $\boldsymbol{\sigma} \cdot \boldsymbol{n}$ in [1.6], it has been proved in [XIA 06] that it can also be guaranteed by the application of [1.8] within the displacement-based FEA framework. The average stress field can be evaluated as the ratio of resultant traction forces on the boundary surfaces to the corresponding areas of the boundary surfaces [XIA 03]

$$\bar{\boldsymbol{\sigma}}_{ij} = \frac{\Delta \boldsymbol{y}_j}{|V|} \int_{S_j} \boldsymbol{\sigma}_{ij} \mathrm{d}S_j = \frac{(P_i)_j}{S_j}, \quad (\text{no summation over}\, j), \qquad [1.9]$$

where S_j is the area of the boundary surface that is oriented perpendicular to the jth direction and $(P_i)_j$ is the resultant traction force acting in the ith direction on boundary surface S_j.

1.3. Finite element discretization

Within the context of FEA, the Voigt notation is employed and the effective stress tensor $\bar{\boldsymbol{\sigma}}$ and strain tensor $\bar{\boldsymbol{\varepsilon}}$ are represented by six-dimensional vectors

$$\begin{cases} \bar{\boldsymbol{\sigma}} \equiv (\bar{\sigma}_{11}, \bar{\sigma}_{22}, \bar{\sigma}_{33}, \bar{\sigma}_{23}, \bar{\sigma}_{13}, \bar{\sigma}_{12})^T, \\ \bar{\boldsymbol{\varepsilon}} \equiv (\bar{\varepsilon}_{11}, \bar{\varepsilon}_{22}, \bar{\varepsilon}_{33}, 2\bar{\varepsilon}_{23}, 2\bar{\varepsilon}_{13}, 2\bar{\varepsilon}_{12})^T. \end{cases} \qquad [1.10]$$

As is shown in Figure 1.3, an RVE is attributed to each Gauss integration point. Without loss of generality, nonlinear elasticity is assumed and body forces are neglected. The macroscopic displacement solution is solved using the iterative NR method:

$$\begin{cases} \bar{\mathbf{K}}_{\tan}^{(k)} \Delta \bar{\mathbf{u}}^{(k+1)} = \bar{\mathbf{f}}_{\text{ext}} - \bar{\mathbf{f}}_{\text{int}}^{(k)}(\bar{\mathbf{u}}^{(k)}) \\ \bar{\mathbf{u}}^{(k+1)} = \bar{\mathbf{u}}^{(k)} + \Delta \bar{\mathbf{u}}^{(k+1)}. \end{cases} \quad [1.11]$$

Here, k indicates the current iteration, $\bar{\mathbf{f}}_{\text{ext}}$ and $\bar{\mathbf{f}}_{\text{int}}$ are the vector forms of the macroscale external and internal forces, and $\bar{\mathbf{u}}$ is the macroscale displacement. $\bar{\mathbf{K}}_{\tan}$ is the consistent macroscale tangent stiffness matrix

$$\bar{\mathbf{K}}_{\tan} = \sum_{e=1}^{N_e} \int_{\Omega_e} \bar{\mathbf{B}}^T \bar{\mathbf{C}}^{\tan} \bar{\mathbf{B}} \mathrm{d}\Omega_e, \quad [1.12]$$

in which N_e is the total number of elements in the macroscale design domain and Ω_e denotes the region of element e. $\bar{\mathbf{C}}^{\tan}$ is the macroscale tangent stiffness. The matrix $\bar{\mathbf{B}}$ relates the strain at material point $\boldsymbol{x}$ and the element displacement vector $\bar{\mathbf{u}}_e$ within the considered element

$$\bar{\varepsilon}(\boldsymbol{x}) = \bar{\mathbf{B}}^T(\boldsymbol{x}) \bar{\mathbf{u}}_e. \quad [1.13]$$

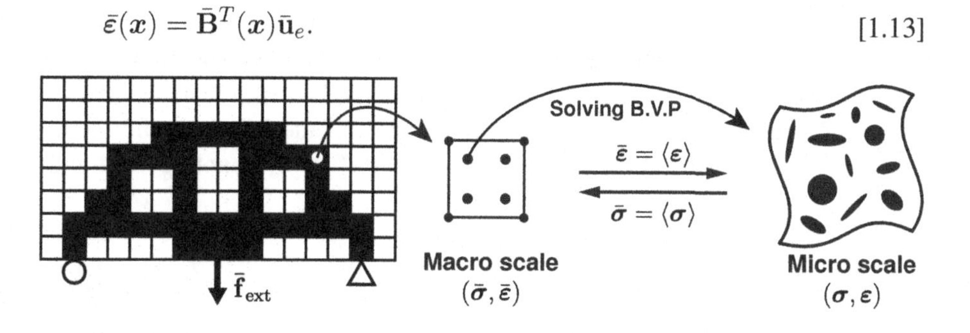

Figure 1.3. *Illustration of the implementation of FE² in the FEA framework [XIA 14b]*

After solving the boundary value problem of the RVE for the given prescribed overall strain load, the macroscopic stress $\bar{\sigma}$ at each integration point can be evaluated via [1.9]. Thereafter, the macroscale internal force vector $\bar{\mathbf{f}}_{\text{int}}$ in [1.11] can be evaluated using the effective stresses returned from the RVE computations

$$\bar{\mathbf{f}}_{\text{int}} = \sum_{e=1}^{N_e} \int_{\Omega_e} \bar{\mathbf{B}}^T \bar{\boldsymbol{\sigma}} \mathrm{d}\Omega_e. \qquad [1.14]$$

The incremental procedure of [1.11] is repeated until the external and internal forces are in balance.

1.4. Topology optimization model

The following is presented with the consideration of the multiscale problem setting presented in section 1.1. Basic definitions of the the topology optimization model are given first in section 1.4.1. Sensitivity analysis with respect to the design variables is derived in section 1.4.2. The BESO variables updating scheme is presented in section 1.4.3.

1.4.1. *Model definitions*

Topology design variables $\boldsymbol{\rho} = (\rho_1, \ldots, \rho_{N_e})^T$ are defined in an elementwise manner, where N_e is the total number of elements in the macroscale design domain. Within the framework of the BESO method (and others), the design variables take values of either 0 or 1, denoting void and solid materials,

$$\rho_e = 0 \text{ or } 1, \; e = 1, \ldots, N_e. \qquad [1.15]$$

whereas in practice an extremely small value $\rho_{\min}$ is attributed to voids to prevent the stiffness matrix singularity.

In linear elastic problems, topology design variables are usually associated with the material's Young's modulus or the element stiffness [SIG 01]. In general nonlinearity when there is no closed-form representation of the material's constitutive law, which is the case for multiscale problems, topology design variables are defined in association with the element internal force vector $\bar{\mathbf{f}}^e_{\text{int}}$ as

$$\bar{\mathbf{f}}^e_{\text{int}} = \rho_e \int_{\Omega_e} \bar{\mathbf{B}}^T \bar{\boldsymbol{\sigma}} \, \mathrm{d}\Omega_e, \qquad [1.16]$$

where the effective stress $\bar{\boldsymbol{\sigma}}$ is computed via the volume averaging relation [1.1]. The microscale stress is determined from an underlying nonlinear microscale equilibrium problem subjected to a prescribed overall strain. In practice, for void elements the microscale solutions can be saved and their effective stresses are set directly to zero.

Two types of design objectives are usually adopted in nonlinear structural designs when the macroscale external force $\bar{\mathbf{f}}_{\text{ext}}$ is imposed, namely the end-compliance

$$f_{\text{c}} = \bar{\mathbf{f}}_{\text{ext}}^T \bar{\mathbf{u}}, \tag{1.17}$$

and the complementary work

$$f_{\text{W}} = \lim_{n \to \infty} \left[\frac{1}{2} \sum_{i=1}^{n} \left(\bar{\mathbf{f}}_{\text{ext}}^{(i)} + \bar{\mathbf{f}}_{\text{ext}}^{(i-1)} \right)^T \Delta \bar{\mathbf{u}}^{(i)} \right], \tag{1.18}$$

where n is the number of load increments. The latter is applied to avoid degenerated topologies, especially when dealing with geometrical nonlinearity [BUH 00]. Without loss of generality, the end-compliance f_{c} is considered. The minimization of the macroscopic structural end-compliance considering a constraint on material volume usage can be formulated as

$$\begin{aligned} \min_{\boldsymbol{\rho}} &: f_{\text{c}}(\boldsymbol{\rho}, \bar{\mathbf{u}}) \\ \text{subject to} &: \bar{\mathbf{r}}(\boldsymbol{\rho}, \bar{\mathbf{u}}) = \mathbf{0} \\ &: V(\boldsymbol{\rho}) = \textstyle\sum \rho_e v_e = V_{\text{req}} \\ &: \rho_e = 0 \text{ or } 1, \ e = 1, \ldots, N_{\text{e}}. \end{aligned} \tag{1.19}$$

Here, $V(\boldsymbol{\rho})$ is the total volume of solid elements, V_{req} is the required volume of solid elements and v_e is the volume of element e. $\bar{\mathbf{u}}$ is the displacement solution at the convergence and $\bar{\mathbf{r}}(\boldsymbol{\rho}, \bar{\mathbf{u}})$ is the residual at the macroscopic scale

$$\bar{\mathbf{r}}(\boldsymbol{\rho}, \bar{\mathbf{u}}) = \bar{\mathbf{f}}_{\text{ext}} - \sum_{e=1}^{N_{\text{e}}} \rho_e \int_{\Omega_e} \bar{\mathbf{B}}^T \bar{\boldsymbol{\sigma}} \, \mathrm{d}\Omega_e. \tag{1.20}$$

1.4.2. *Sensitivity analysis*

To implement topology optimization, sensitivity of the design objective with respect to design variables needs to be provided. The derivation of the sensitivity requires using the adjoint method [BUH 00]. Introducing a vector of Lagrangian multipliers $\boldsymbol{\lambda}$, we may rewrite objective [1.19] in the following form without modifying the objective value

$$f_{\mathrm{c}}^{*}(\boldsymbol{\rho}) = \bar{\mathbf{f}}_{\mathrm{ext}}^{T}\bar{\mathbf{u}} + \boldsymbol{\lambda}^{T}\bar{\mathbf{r}}, \qquad [1.21]$$

where the term $\boldsymbol{\lambda}^{T}\bar{\mathbf{r}}$ equals zero when the equilibrium of [1.20] is achieved, i.e. $f_{\mathrm{c}}^{*} = f_{\mathrm{c}}$.

Note that $\bar{\mathbf{f}}_{\mathrm{ext}}$ is invariant to the variation of design variables $\partial\bar{\mathbf{f}}_{\mathrm{ext}}/\partial\rho_e = \mathbf{0}$ and according to the residual definition [1.20], the derivative of the modified objective function f_{c}^{*} with respect to ρ_e equals

$$\frac{\partial f_{\mathrm{c}}^{*}}{\partial\rho_e} = \bar{\mathbf{f}}_{\mathrm{ext}}^{T}\frac{\partial\bar{\mathbf{u}}}{\partial\rho_e} + \boldsymbol{\lambda}^{T}\left(\frac{\partial\bar{\mathbf{r}}}{\partial\bar{\mathbf{u}}}\frac{\partial\bar{\mathbf{u}}}{\partial\rho_e} - \int_{\Omega_e}\bar{\mathbf{B}}^{T}\bar{\boldsymbol{\sigma}}\,\mathrm{d}\Omega_e\right). \qquad [1.22]$$

With the purpose of eliminating $\partial\bar{\mathbf{u}}/\partial\rho_e$, regrouping the terms with $\partial\bar{\mathbf{u}}/\partial\rho_e$ in [1.22] yields

$$\frac{\partial f_{\mathrm{c}}^{*}}{\partial\rho_e} = \left(\bar{\mathbf{f}}_{\mathrm{ext}}^{T} - \boldsymbol{\lambda}^{T}\bar{\mathbf{K}}_{\mathrm{tan}}\right)\frac{\partial\bar{\mathbf{u}}}{\partial\rho_e} - \boldsymbol{\lambda}^{T}\int_{\Omega_e}\bar{\mathbf{B}}^{T}\bar{\boldsymbol{\sigma}}\,\mathrm{d}\Omega_e, \qquad [1.23]$$

in which

$$\bar{\mathbf{K}}_{\mathrm{tan}} = -\frac{\partial\bar{\mathbf{r}}}{\partial\bar{\mathbf{u}}} \qquad [1.24]$$

is the macroscale tangent stiffness matrix. Recall the symmetry of $\bar{\mathbf{K}}_{\mathrm{tan}}$, i.e. $\bar{\mathbf{K}}_{\mathrm{tan}}^{T} = \bar{\mathbf{K}}_{\mathrm{tan}}$, the first term of the right-hand side of [1.23] can be eliminated by imposing $\boldsymbol{\lambda}$ as the solution of the following self-adjoint problem

$$\bar{\mathbf{K}}_{\mathrm{tan}}\boldsymbol{\lambda} = \bar{\mathbf{f}}_{\mathrm{ext}}, \qquad [1.25]$$

and yields

$$\frac{\partial f_{\mathrm{c}}}{\partial\rho_e} = \frac{\partial f_{\mathrm{c}}^{*}}{\partial\rho_e} = -\boldsymbol{\lambda}^{T}\int_{\Omega_e}\bar{\mathbf{B}}^{T}\bar{\boldsymbol{\sigma}}\,\mathrm{d}\Omega_e. \qquad [1.26]$$

1.4.3. *BESO updating scheme*

In the BESO method [HUA 10b], an evolutionary ratio c_{er} is defined to determine the required volume of material usage at each design iteration following

$$V^{(l)} = \max\left\{V_{\text{req}}, (1 - c_{\text{er}})V^{(l-1)}\right\}, \qquad [1.27]$$

in which $V^{(l)}$ and $V^{(l-1)}$ denote the required volumes of the solid at the current (lth) iteration and the previous iteration, respectively. Note that in general the volume of the solid of the structure decreases iteratively until the required volume V_{req} is achieved.

At each design iteration, the sensitivity numbers that denote the relative ranking of the elemental sensitivities are used to determine material removal and addition. The sensitivity number for the considered objective is defined as the opposite of the sensitivity divided by the element volume

$$\alpha_e = -\frac{\partial f_{\text{c}}}{\partial \rho_e}\frac{1}{v_e}. \qquad [1.28]$$

Note that the division by element volumes can be omitted when uniform mesh is used.

In order to avoid mesh dependency and checkerboard pattern, sensitivity numbers are smoothed by means of a filtering scheme [LI 01, SIG 01]

$$\alpha_e = \frac{\sum_{j=1}^{N_{\text{e}}} w_{ej}\alpha_j}{\sum_{j=1}^{N_{\text{e}}} w_{ej}}, \qquad [1.29]$$

where w_{ej} is a linear weight factor

$$w_{ej} = \max(0, r_{\min} - \Delta(e,j)), \qquad [1.30]$$

determined according to the prescribed filter radius $r_{\min}$ and the element center-to-center distance $\Delta(e,j)$. A schematic illustration of the filtering scheme is shown in Figure 1.4, where a checkerboard field is filtered with $r_{\min} = 1.5$ and $r_{\min} = 3$ times of element length l_e, respectively. It can be seen that the concerned field is smoothed by the filter scheme, for which reason the sensitivity numbers of void elements can be naturally obtained. By this scheme, void elements neighboring the regions of high sensitivity numbers have a higher potential to be recovered in the next iteration.

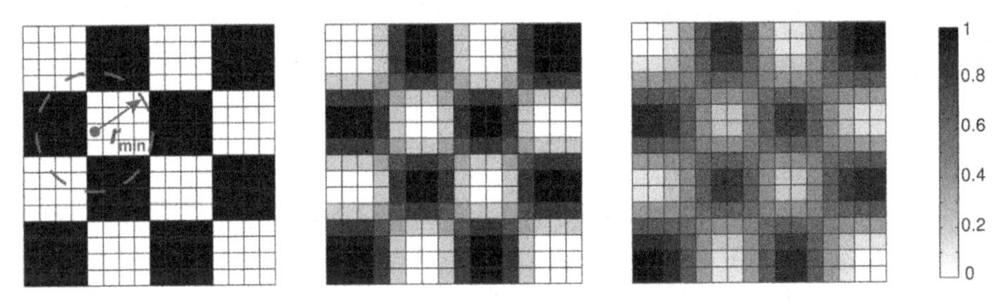

Figure 1.4. *A checkerboard field and filtered fields ($r_{\min} = 1.5l_e$ and $3l_e$)*

It has been examined that the topology and the objective may encounter difficulties in convergence due to the discrete nature of the BESO material model. To improve the convergence of the solution, we may simply average the current sensitivity numbers with their historical information [HUA 07a]

$$\alpha_e^{(l)} \leftarrow (\alpha_e^{(l)} + \alpha_e^{(l-1)})/2. \quad [1.31]$$

For variable updating, a threshold of sensitivity number α_{th} is determined by means of a bisection algorithm from all sensitivity numbers satisfying the target volume at the current design iteration [HUA 09]. The design variables are updated according to

$$\rho_e = \max\left\{\rho_{\min}, \text{sign}(\alpha_e - \alpha_{\text{th}})\right\}, \quad [1.32]$$

which means solids will be switched to voids if α_e is lower than α_{th}, accordingly voids will be switched back to solids when α_e is higher than α_{th}. The evolutionary design process stops when the objective value or the structural topology reaches convergence.

1.5. Multiscale design framework

The general algorithm for the multiscale design framework consisting of topology optimization and FE^2 is outlined in algorithm 1.1. Generally speaking, there exist three layers in this framework. The very outer layer is the optimization that loops until the design solution has reached the design convergence δ_{opt}. The middle and inner layers are the nested macroscopic and microscopic boundary value problems, i.e. FE^2. In the middle layer, the iterative NR solution procedure at the macroscopic scale for each load increment loops until the residual attains the convergence criterion δ_{f}. The inner layer is implicitly contained through solving the RVE problems and each of them is a complete nonlinear equilibrium problem subjected to the PBC according to the associated effective strain value $\bar{\varepsilon}$. From this framework illustration, we can clearly view the multiscale dilemma in terms of heavy computational burden and enormous storage requirements.

Algorithm 1.1. Multiscale design framework

1: Initial $\boldsymbol{\rho}_0$;
2: **while** $\|\boldsymbol{\rho}_{i+1} - \boldsymbol{\rho}_i\| > \delta_{\text{opt}}$ $\{i++\}$ **do**
3:     **loop** over all macro time steps $\{j++\}$
4:         **while** $\|\bar{\mathbf{f}}_{\text{ext}} - \bar{\mathbf{f}}_{\text{int}}\| > \delta_{\text{f}}$ $\{k++\}$ **do**
5:             **loop** over macroscale Gauss points
6:                 compute the effective strain $\bar{\varepsilon}$;
7:                 define PBC on the associated RVE upon $\bar{\varepsilon}$;
8:                 compute $\boldsymbol{\sigma}$ from solving the RVE problem;
9:                 compute the effective stress through $\bar{\boldsymbol{\sigma}} = \langle \boldsymbol{\sigma} \rangle$;
10:                compute the effective tangent stiffness $\bar{\mathbf{C}}^{\text{tan}}$;
11:            **end loop**
12:            NR update: $\bar{\mathbf{K}}_{\text{tan}} \Delta \bar{\mathbf{u}} = \bar{\mathbf{f}}_{\text{ext}} - \sum \rho_e \int_{\Omega_e} \bar{\mathbf{B}}^T \bar{\boldsymbol{\sigma}} \mathrm{d}\Omega_e$;
13:        **end while**
14:    **end loop**
15:    compute f_{c} and sensitivities $\partial f_{\text{c}} / \partial \boldsymbol{\rho}$;
16:    update $\boldsymbol{\rho}$ using the BESO scheme;
17: **end while**
18: **return** $\boldsymbol{\rho}$.

1.6. Numerical examples

This section presents two test examples. For the purpose of comparison, topology optimization design for a homogeneous structure in the case of nonlinear elasticity is first presented in section 1.6.1. The second example in section 1.6.2 deals with the design of a heterogeneous structure made of short-fiber reinforced composite, which is constituted by a nonlinear elastic matrix and rigid elastic fibers.

1.6.1. *Design of a nonlinear homogeneous structure*

A cantilever discretized into 100×50 square-shaped bilinear elements is considered as shown in Figure 1.5. Element dimensions are 1×1 mm^2 and the structure is assumed in plane strain condition. The left end of the cantilever is fixed and an external force is applied on the middle point of the right edge. In regard to topology optimization, inefficient or redundant materials are gradually removed according to the sensitivity ranking from an initial full solid design in an evolutionary rate of $c_{\text{er}} = 2\%$. Sensitivity numbers are filtered within a local zone controlled by a filter radius $r_{\min} = 6$ mm. The constraint on the volume fraction of solid is set to 60%. For the purpose of comparison, the linear elastic topology design result obtained using the same parameter set is also given in Figure 1.5.

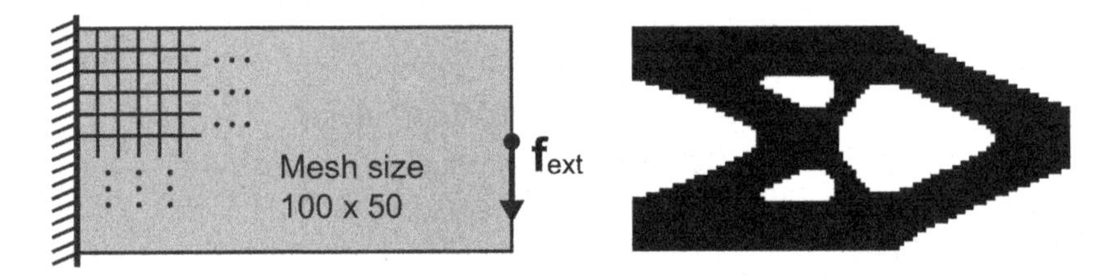

Figure 1.5. *Illustration of a cantilever and its linear topology design of 60% volume fraction*

In regard to nonlinear design, the present work is limited to nonlinear elasticity subjected to small deformations. The considered nonlinear constitutive behavior is governed by an isotropic compressible potential of the form

$$w(\varepsilon) = \frac{9}{2}\kappa\varepsilon_m + \frac{\varepsilon_0\sigma_0}{1+m}\left(\frac{\varepsilon_{eq}}{\varepsilon_0}\right)^{1+m}. \qquad [1.33]$$

Here, κ denotes the bulk modulus, $\varepsilon_m = Tr(\varepsilon)/3$ is the hydrostatic strain and ε_{eq} is the equivalent strain defined by $\varepsilon_{eq} = \sqrt{2\varepsilon_d : \varepsilon_d/3}$ with $\varepsilon_d = \varepsilon -$

$\varepsilon_m \mathbf{1}$ and $\mathbf{1}$ being the second-order identity tensor. m is the strain-hardening parameter such that $0 \leq m \leq 1$. σ_0 and ε_0 are the flow stress and reference strain, respectively. The stress–strain relationship is provided by:

$$\boldsymbol{\sigma} = \frac{\partial w(\boldsymbol{\varepsilon})}{\partial \boldsymbol{\varepsilon}} = \kappa Tr(\boldsymbol{\varepsilon})\mathbf{1} + \frac{2}{3}\frac{\sigma_0}{\varepsilon_0}\left(\frac{\varepsilon_{eq}}{\varepsilon_0}\right)^{m-1} \boldsymbol{\varepsilon}_d. \qquad [1.34]$$

This is a commonly used constitutive model for the representation of a number of nonlinear mechanical phenomena [DUV 84, NIX 98, YVO 09]. In particular, the cases $m = 0$ and $m = 1$ correspond to perfectly rigid plastic and linearly elastic materials, respectively.

The following numerical parameters are chosen for the current test case: $m = 0.5$, $\kappa = 20$ MPa, $\sigma_0 = 1$ MPa and $\varepsilon_0 = 1$. Nonlinear topology optimization designs are carried out for three different load forces 0.01, 0.2 and 0.4 N and the corresponding topology design results are shown in Figures 1.6(a)–(c), respectively. The nonlinear design algorithm gives an almost identical topology design result as in linear elasticity when the load force is small, as can be viewed from Figure 1.6(a) and the linear topology solution in Figure 1.5. When the load force increases, the topology design result varies in response to the load force value as can be observed from Figures 1.6(b) and (c) for $\mathbf{f}_{\text{ext}} = 0.2$ N and 0.4 N, respectively. From Figure 1.6, we observe that material moves towards the left end of the cantilever to resist the increasing load force. The equivalent stress fields of the three topologies are also given in Figure 1.6 on exaggeratedly deformed meshes. For the purpose of illustrations, the elements neighboring the loading tip with high stress concentration are removed from the stress field plots.

1.6.2. *Design of a nonlinear heterogeneous structure*

In this example, a two-scale cantilever structure made of periodically patterned anisotropic short-fiber reinforced composite as shown in Figure 1.7 is to be designed. Following [YVO 09], both the matrix (phase 1) and the fibers (phase 2) are assumed to be isotropic and compressible materials characterized by the governing potential of [1.33]. The matrix material is highly nonlinear: $m^{(1)} = 0.5$, $\kappa^{(1)} = 20$ MPa, $\sigma_0^{(1)} = 1$ MPa and $\varepsilon_0^{(1)} = 1$. The fibers are assumed to be linear elastic and much more rigid than the matrix: $m^{(2)} = 1$, $\kappa^{(2)} = 20$ MPa, $\sigma_0^{(2)} = 1,000$ MPa and $\varepsilon_0^{(2)} = 1$. The RVE

(Figure 1.5) is discretized into 20 $\times$ 20 square bilinear elements. The equivalent stress fields within the RVE in cases of biaxial stretching and uniaxial stretching combined with shear are shown in Figure 1.8.

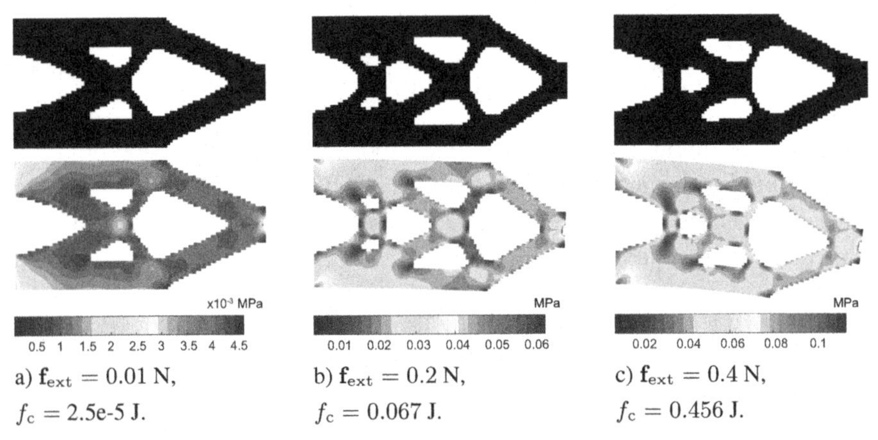

a) $\mathbf{f}_{\text{ext}} = 0.01$ N, $f_c = 2.5\text{e-}5$ J. b) $\mathbf{f}_{\text{ext}} = 0.2$ N, $f_c = 0.067$ J. c) $\mathbf{f}_{\text{ext}} = 0.4$ N, $f_c = 0.456$ J.

Figure 1.6. *Nonlinear topology designs for three different loading cases and their equivalent stress fields (deformation exaggerated 10 times). For a color version of the figure, see www.iste.co.uk/xia/topology.zip*

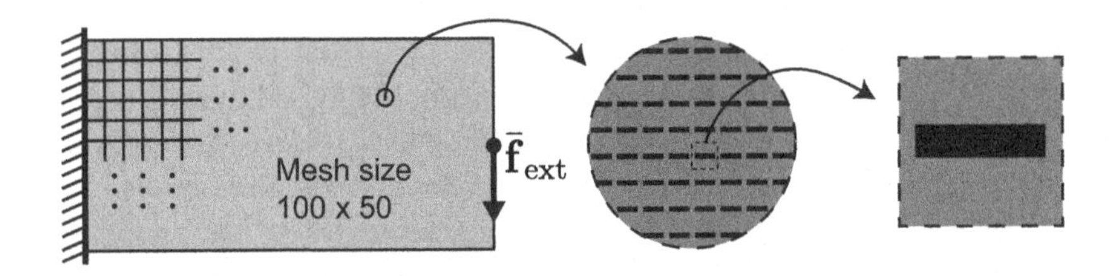

Figure 1.7. *Illustration of a two-scale cantilever made up of periodically patterned short-fiber reinforced composite*

Topology optimization is carried out for the macroscale structure with the same BESO parameters that are used in the previous example in section 1.6.1, i.e. the evolutionary rate $c_{\text{er}} = 2\%$, the filter radius $r_{\min} = 6$ mm, the volume fraction constraint is 60%. It important to emphasize that it requires solving $4 \times 100 \times 50$ (4 Gauss integration points for each element) nonlinear RVE boundary value problems for each iteration of each load increment. This number would decrease progressively with iterations as the removed elements are no longer evaluated for the structural response.

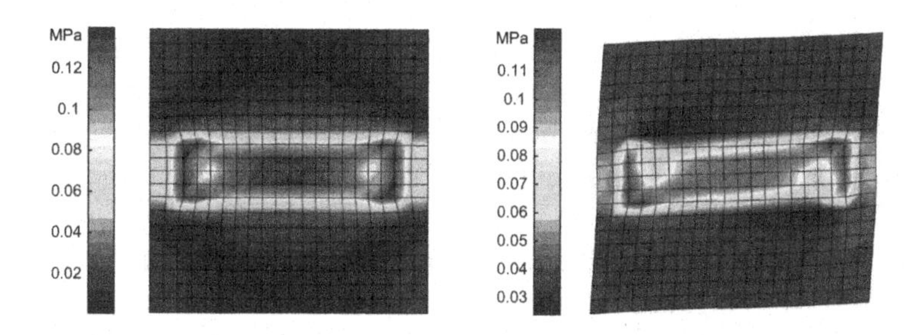

Figure 1.8. *Equivalent stress fields (deformation exaggerated 50 times) of the short-fiber reinforced RVE for biaxial stretching (left, $\bar{\varepsilon}_{11} = \bar{\varepsilon}_{22} = 0.002, \bar{\varepsilon}_{12} = 0$) and uniaxial stretching with shear (right, $\bar{\varepsilon}_{11} = 0.001, \bar{\varepsilon}_{22} = 0, \bar{\varepsilon}_{12} = 0.002$). For a color version of the figure, see www.iste.co.uk/xia/topology.zip*

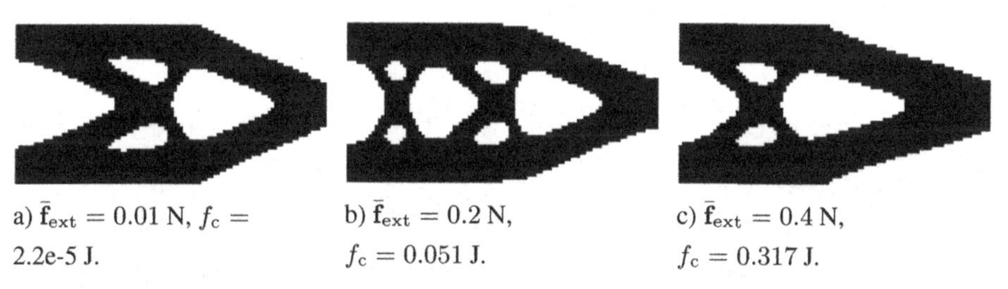

a) $\bar{\mathbf{f}}_{\mathrm{ext}} = 0.01$ N, $f_c =$ 2.2e-5 J.

b) $\bar{\mathbf{f}}_{\mathrm{ext}} = 0.2$ N, $f_c = 0.051$ J.

c) $\bar{\mathbf{f}}_{\mathrm{ext}} = 0.4$ N, $f_c = 0.317$ J.

Figure 1.9. *Nonlinear topology designs for the two-scale heterogeneous structure*

For the purpose of comparison, designs are also carried out for the same three load forces, i.e. 0.01, 0.2 and 0.4 N, as considered in section 1.6.1 and the corresponding design results are shown in Figures 1.9(a)–(c). The topology shown in Figure 1.9 is similar to the topologies of Figures 1.5 and 1.6(a), indicating that an external force load at the level of 0.01 N does not result in much difference in the design results. However, when the external load is increased to 0.2 and 0.4 N, we can observe obvious topological differences between the design results shown in Figures 1.9(b) and (c) and 1.6(b) and (c), respectively, which are due to the existence of the reinforcing fibers. The presence of fibers also results in lower end-compliance values, i.e. increased stiffness, of the design results (Figures 1.9 and 1.6).

The equivalent stress field of the topology solution in Figure 1.9(b) is given in Figure 1.10 together with the equivalent stress fields of the RVEs at several selected points. The elements neighboring the loading tip with high stress concentration are removed from the macroscale field plot for the purpose of illustration. From the deformed RVEs shown in Figure 1.10, we can observe that the RVEs at points A and D are under compression, the RVE at point B is under tension, and the RVE at point C is subjected to a mechanical shear state, which are in agreement with their macroscale deformation states. We may also note from the stress fields that the presence of fibers results in higher stress concentrations at the interface of the matrix and the fibers. The higher stress concentrations are responsible for the initial material failure or crack at the microscopic scale, which cannot be detected when using the conventional monoscale fracture analysis [COE 12]. At the microscopic scale, there is also a potential application in stress-related topological designs [DUY 98, LE 10, GUO 11, BRU 12, ZHA 13, CAI 14, CAI 15] by including a stress constraint to limit the maximum stress at the microscopic scale.

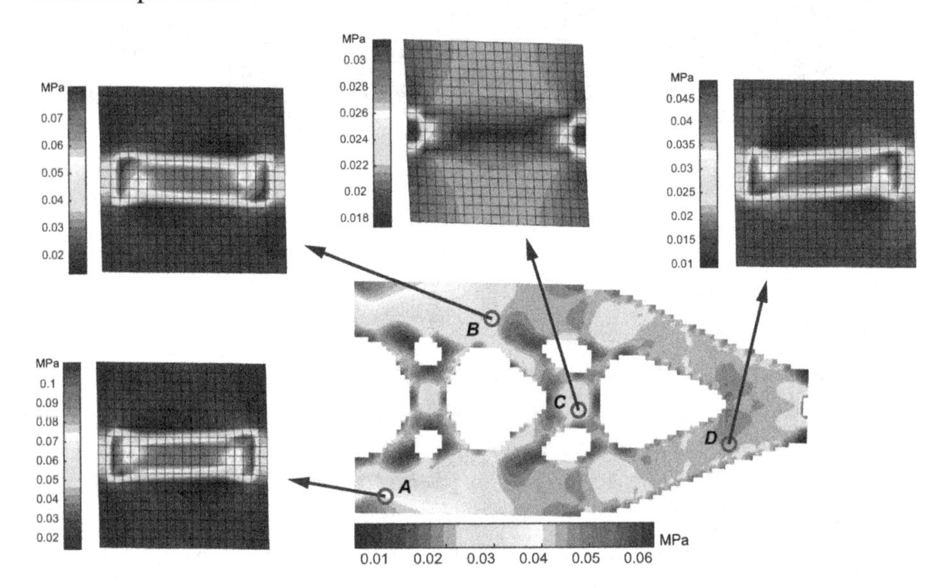

Figure 1.10. *The equivalent stress fields of case b in Figure 1.9 for the macrostructure (deformation exaggerated 10 times) and for the micro-RVEs at selected points (deformation exaggerated 50 times). For a color version of the figure, see www.iste.co.uk/xia/topology.zip*

1.7. Concluding remarks

In this chapter, we have developed a multiscale design framework through a synthesis of topology optimization and multiscale modeling. In contrast to the conventional nonlinear design of homogeneous structures, this design framework provides an automatic design tool by which the considered material models can be governed directly by the realistic microstructural geometry and the microscopic constitutive models.

The main difficulty of such design lies on the unbearable computational burden and data storage requirement due to multiple realizations of FE^2 computation. In Chapters 2 and 3, we will present model reduction strategies to alleviate the computational burden for the cases of nonlinear elasticity and elastoviscoplasticity, respectively.

By introducing topology design variables to the microscopic scale, the currently developed multiscale design framework can also be extended to perform simultaneous topology optimization of structure and materials at the two scale. This subject will be presented in Chapters 4 and 5.

2

POD-based Adaptive Surrogate for the Design of Multiscale Structures

This chapter is dedicated to alleviating the multiscale dilemma in terms of heavy computational burden of the multiscale design framework discussed in Chapter 1. We note that the optimization design process requires multiple design loops involving similar or even repeated computations at the microscopic scale, which potentially suits the surrogate learning process. We therefore develop an adaptive surrogate model in this chapter using snapshot POD and diffuse approximation for the solution of the microscopic problems.

The proposed method is a non-intrusive reduction approach, which is an alternative to the intrusive approach [YVO 07]. The reduced basis is extracted by means of snapshot POD where the displacement solutions at the microscopic scale are treated as snapshots. The surrogate model is constructed using diffuse approximation [NAY 92], a variant of moving least squares [LAN 81]. The surrogate is constructed in an on-line manner: initially built during the first optimization design iteration is then updated in the following design iterations.

With the purpose of demonstrating the independence of the multiscale design framework developed in Chapter 1 to the applied topology optimization model, in this chapter we choose a discrete level set method [CHA 10] instead of the BESO method to perform topology optimization. Similar to the BESO method, the discrete version of the level set method can straightforwardly link the RVE to the solid region of the structure without defining a pseudo-relationship between the intermediate values and the considered RVE.

The remainder of the chapter is organized in the following manner: we review first in section 2.1 the multiscale design framework using a discrete level set model for topology optimization. Section 2.2 gives the POD-based adaptive surrogate constructed using snapshot POD and diffuse approximation. Section 2.3 showcases the performance of the surrogate model using two examples. Concluding remarks are given in section 2.4

2.1. Multiscale design framework

The following framework is presented with the consideration of the multiscale problem setting (FE^2) presented in section 1.1. Matrix and vector forms are applied in accordance with the finite element method. The macroscopic stress $\bar{\sigma}$ and strain $\bar{\varepsilon}$ represent their vector forms following the Voigt notation [1.10].

2.1.1. *Discrete level set model*

An initial level set function $\varphi(x, t_0)$ is constructed as a signed distance function upon the discretized initial structural topology as shown in Figure 2.1

$$\begin{cases} \varphi(x_e, t_0) < 0 \text{ if } \rho_e = 1 \\ \varphi(x_e, t_0) > 0 \text{ if } \rho_e = 0, \end{cases} \qquad [2.1]$$

where x_e denotes the center of the eth element and ρ_e is its pseudo-density. By constructing $\varphi(x_e, t_0) \neq 0$, $\rho_e = 1$ or 0 indicate element e is occupied by solid or void material correspondingly in the sense of discrete topology optimization design, and no intermediate value is attributed to ρ_e.

The initialized level set function $\varphi(x, t_0)$ is then updated to $\varphi(x, t)$ corresponding to a new structural topology by solving the "Hamilton–Jacobi" evolution equation

$$\frac{\partial \varphi(x,t)}{\partial t} + \vartheta_n |\nabla \varphi(x,t)| = 0, \qquad [2.2]$$

where t is a pseudo-time defined corresponding to different optimization iterations. The normal velocity field ϑ_n determines the geometric motion of the boundary of the structure and is chosen based on the shape derivative of the design objective. The updated level set function $\varphi(x, t)$ is then mapped to the discretized design domain through

$$\rho_e = \begin{cases} 1, \text{ if } \varphi(x_e, t) \leq 0, \\ 0, \text{ if } \varphi(x_e, t) > 0. \end{cases} \quad [2.3]$$

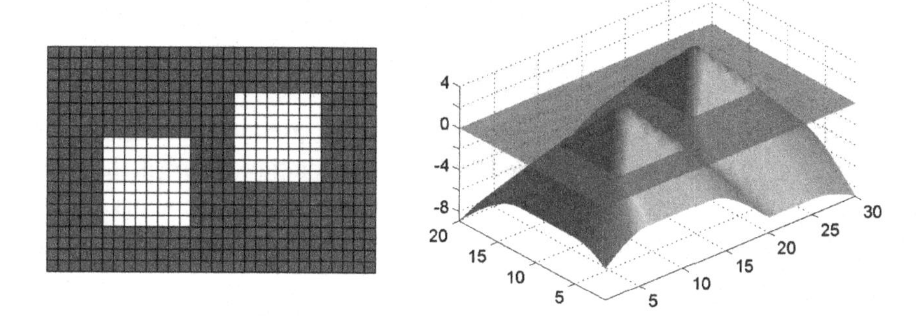

Figure 2.1. *A rectangular plate with two square holes and its discretized level set function*

Note that, in practice, in order to prevent the singularity of the stiffness matrix, a small value is attributed to ρ_e to denote void elements.

2.1.2. *Optimization model*

With the aforementioned definitions, we have in fact introduced the concept of element pseudo-density in terms of level set function $\rho(\varphi)$ to perform the discretized topology optimization. At the macroscopic scale, material constitutive behavior is implicitly given in terms of the effective stress–strain relationship evaluated by the FE^2 method. In analogy to [1.16], topology design variables are defined in association with the element internal force vector $\bar{\mathbf{f}}_{\text{int}}^e$

$$\bar{\mathbf{f}}_{\text{int}}^e = \rho_e \int_{\Omega_e} \bar{\mathbf{B}}^T \bar{\boldsymbol{\sigma}} \, \mathrm{d}\Omega_e, \quad [2.4]$$

where $\bar{\boldsymbol{\sigma}}$ is the effective stress computed from the microscale stress, which is determined through solving the corresponding RVE problem.

The same optimization model as [1.19] is adopted for the minimization of the macroscopic structural end-compliance subjected to one constraint on material volume usage:

$$\begin{aligned} \min_{\boldsymbol{\rho}(\varphi)} &: f_c = \bar{\mathbf{f}}_{\text{ext}}^T \bar{\mathbf{u}} \\ \text{subject to} &: \bar{\mathbf{r}}(\boldsymbol{\rho}, \bar{\mathbf{u}}) = \mathbf{0} \\ &: V(\boldsymbol{\rho}) = \sum \rho_e v_e = V_{\text{req}} \\ &: \rho_e = 0 \text{ or } 1, \forall e = 1, \ldots, N_{\text{e}}, \end{aligned} \qquad [2.5]$$

where $\boldsymbol{\rho} = (\rho_1, \ldots, \rho_{N_{\text{e}}})$ is the vector of the element pseudo-densities. In the following, we will denote $\boldsymbol{\rho}(\varphi)$ by $\boldsymbol{\rho}$ to reduce the notation. $V(\boldsymbol{\rho})$ is the total volume of solid elements, V_{req} is the required volume of solid elements and v_e is the volume of element e. $\bar{\mathbf{u}}$ is the displacement solution at the convergence and $\bar{\mathbf{r}}(\boldsymbol{\rho}, \bar{\mathbf{u}})$ is the macroscale residual.

The augmented Lagrangian method is applied to convert the original constrained optimization problem [2.5] into an unconstrained problem as presented in [BEL 03, LUO 08, CHA 10]

$$L = f_c + \lambda^{(k)}(V(\boldsymbol{\rho}) - V_{\text{req}}) + \frac{1}{2\Lambda^{(k)}}(V(\boldsymbol{\rho}) - V_{\text{req}})^2, \qquad [2.6]$$

where $\lambda^{(k)}$ is the Lagrangian multiplier and $\Lambda^{(k)}$ is the penalty parameter updated iteratively with the optimization iteration k using the scheme [LUO 08]:

$$\lambda^{(k+1)} = \lambda^{(k)} + \frac{1}{\Lambda^{(k)}}(V(\boldsymbol{\rho}) - V_{\text{req}}), \;\; \Lambda^{(k+1)} = \alpha\Lambda^{(k)}, \qquad [2.7]$$

where $\alpha \in (0, 1)$ is a parameter to be fixed by the user (see section 2.3). The initial values of λ and Λ are decided according to the physical responses considered [LUO 08, CHA 10].

In order to update the level set function $\varphi(x, t)$ and therefore the structural topology $\boldsymbol{\rho}(\varphi)$, the normal velocity ϑ_n needs to be determined in [2.2]. Conventionally, ϑ_n is chosen as a descent direction for the Lagrangian L

using its shape derivative [WAN 03, ALL 04b]. The normal velocity ϑ_n within an element e at design iteration k is derived as

$$\vartheta_n|_e = -\frac{\partial L}{\partial \rho_e} = -\frac{\partial f_c}{\partial \rho_e} - \lambda^{(k)} v_e - \frac{v_e}{\Lambda^{(k)}}(V(\boldsymbol{\rho}) - V_{\text{req}}), \qquad [2.8]$$

with

$$\frac{\partial f_c}{\partial \rho_e} = -\left(\bar{\mathbf{K}}_{\tan}^{-1} \bar{\mathbf{f}}_{\text{ext}}\right)^T \int_{\Omega_e} \bar{\mathbf{B}}^T \bar{\boldsymbol{\sigma}} \, \mathrm{d}\Omega_e, \qquad [2.9]$$

as is derived in section 1.4.2.

It is worth noting that the standard evolution equation of [2.2] cannot nucleate new void regions during the optimization process [ALL 04b]. An additional forcing term based on the topological derivative of the design objective can be integrated into [2.2] to nucleate new holes within the structure [BUR 04]. In this work, we follow the initial algorithm [WAN 03, ALL 04b], for which a topology with "sufficient" holes is initially defined. The initial layout of holes is arbitrarily fixed. Though new holes cannot be nucleated, the initially defined holes can merge and split during the design, providing sufficient freedom for topological design or at least sufficient enough for practical applications.

2.2. POD-based adaptive surrogate

The adaptive surrogate model is constructed through coupling snapshot POD and diffuse approximation. The first level of reduction is achieved by snapshot POD, allowing us to expand a displacement field as a linear combination of the truncated modes. Second, the surrogate model based on diffuse approximation is built to express the POD projection coefficients as functions of the effective strain tensors.

2.2.1. *Snapshot POD*

We consider a D-dimensional ($D = 2$ or 3) RVE of N points subjected to a time-dependent loading $\bar{\varepsilon}(t)$ during a time interval $I = [0, T]$ discretized by M instants $\{t_1, t_2, \ldots, t_M\}$. Let $\mathbf{u}^{(i)} \in \mathbb{R}^{DN}$ denote the DN-dimensional nodal displacement vector called snapshot recorded at the instant $t_i \in I$. The reduced-order displacement vector $\tilde{\mathbf{u}}(t) \in \mathbb{R}^{DN}$ may be written as [YVO 07]

$$\tilde{\mathbf{u}}(t) = \boldsymbol{\mu} + \sum_{i=1}^{m} \boldsymbol{\phi}_i \alpha_i(\bar{\varepsilon}(t)), \tag{2.10}$$

where $m \ll \min(M, DN)$, $\boldsymbol{\mu} = 1/M \sum_{i=1}^{M} \mathbf{u}^{(i)}$ is the average displacement, $\boldsymbol{\phi}_i \in \mathbb{R}^{DN}$ are constant vectors and coefficients $\alpha_i(\bar{\varepsilon}(t))$ are scalar functions of pseudo-time t. $\boldsymbol{\phi}_i$ are the eigenvectors of the eigenvalue problem

$$\mathbf{C_u}\boldsymbol{\phi}_i = \lambda_i \boldsymbol{\phi}_i, \tag{2.11}$$

where $\mathbf{C_u}$ is the covariance matrix

$$\mathbf{C_u} = \mathbf{D_u}\mathbf{D_u}^T, \tag{2.12}$$

where $\mathbf{D_u}$ is the deviation matrix of dimensionality $DN \times M$ composed of centered nodal displacement vectors as columns

$$\mathbf{D_u} = \{\mathbf{u}^{(1)} - \boldsymbol{\mu}, \ldots, \mathbf{u}^{(M)} - \boldsymbol{\mu}\}. \tag{2.13}$$

The coefficients $\alpha_j^{(i)}$ for the ith snapshot $\mathbf{u}^{(i)}$ are the projections of this snapshot on the basis

$$\alpha_j^{(i)} = \boldsymbol{\phi}_j^T \mathbf{u}^{(i)}. \tag{2.14}$$

The size of the truncated basis m is chosen in consideration of the reconstruction error of $\mathbf{D_u}$

$$\epsilon = 1 - \frac{\sum_{i=1}^{m} \lambda_i}{\sum_{j=1}^{M} \lambda_j} < \delta, \tag{2.15}$$

where δ is a prescribed tolerance.

2.2.2. *Interpolation of the projection coefficients*

The surrogate model for each of the projection coefficients α_i, $i = 1, \ldots, m$, with respect to average stain $\bar{\varepsilon}$ in [2.10] is constructed using the method of diffuse approximation [BRE 98], an extension of moving least squares

$$\alpha_i(\bar{\varepsilon}) \approx \tilde{\alpha}_i(\bar{\varepsilon}) = \mathbf{p}^T(\bar{\varepsilon})\mathbf{a}^{(i)}(\bar{\varepsilon}), \quad [2.16]$$

where $\mathbf{p} = [p_1, p_2, \ldots]^T$ is the polynomial basis vector. In the 2D case, the polynomial basis vector expressed in terms of the macroscale effective strain is

$$\mathbf{p} = [1, \bar{\varepsilon}_{11}, \bar{\varepsilon}_{22}, \bar{\varepsilon}_{12}, \ldots]^T. \quad [2.17]$$

The superscript of the vector of coefficients $\mathbf{a}^{(i)} = [a_1^{(i)}, a_2^{(i)}, \ldots]^T$ indicates that the coefficient vector is dependent on the projection coefficients and they are the minimizers of functional defined by

$$J(\mathbf{a}^{(i)}) = \frac{1}{2}\sum_{k=1}^{M} w_k \left(\mathbf{p}^T(\bar{\varepsilon})\mathbf{a}^{(i)}(\bar{\varepsilon}) - \alpha_i(\bar{\varepsilon}(t_k))\right)^2, \quad [2.18]$$

in which w_k are the weights depending on the Euclidean distance between $\bar{\varepsilon}$ and $\bar{\varepsilon}(t_k)$

$$w_k = w_{\text{ref}}\left(\frac{\text{dist}\,(\bar{\varepsilon}, \bar{\varepsilon}(t_k))}{r_{\text{diff}}}\right), \quad [2.19]$$

where r_{diff} is a radius defining the local influence zone. w_{ref} is chosen here as a piecewise cubic spline expressed as [BRE 04]

$$w_{\text{ref}}(s) = \begin{cases} 1 - 3s^2 + 2s^3, \text{ if } 0 \leq s \leq 1, \\ 0, \qquad\qquad\quad\ \text{if } s \geq 1. \end{cases} \quad [2.20]$$

2.2.3. *Adaptive POD-based surrogate*

An illustrative flowchart of the approximation procedure is given in Figure 2.2. With a given admissible value of macroscale strain $\bar{\varepsilon}^*$, the corresponding approximated POD projection coefficients from $\tilde{\alpha}_1$ to $\tilde{\alpha}_m$ are locally interpolated using diffuse approximation. Substituting the approximated coefficients into [2.10], we have the reduced order solution of the displacement filed

$$\tilde{\mathbf{u}}(\bar{\varepsilon}^*) = \boldsymbol{\mu} + \boldsymbol{\Phi}\tilde{\boldsymbol{\alpha}}(\bar{\varepsilon}^*), \quad [2.21]$$

with $\boldsymbol{\Phi} = \{\phi_1, \ldots, \phi_m\}$ the truncated POD basis.

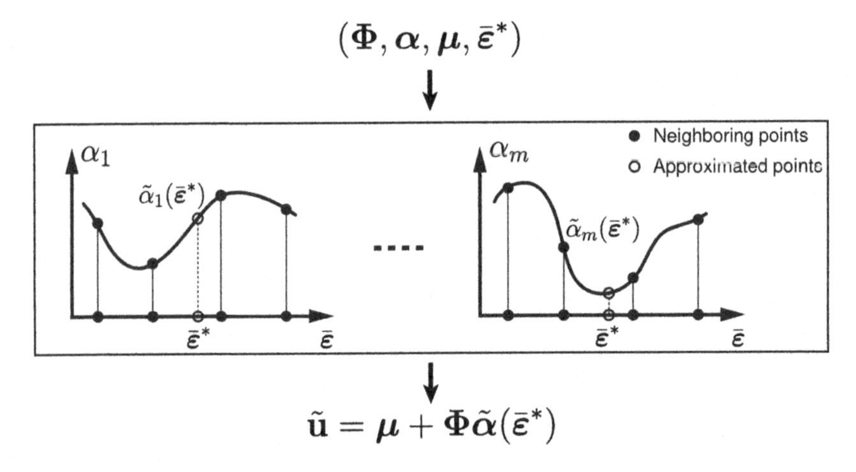

Figure 2.2. *Illustration of the approximation procedure of the surrogate model*

The surrogate is applied to provide microscale RVE solutions instead of performing FEA. A detailed solution scheme of the microscale RVE problem is given in algorithm 2.1. Computations during the first time step of the first optimization iteration are performed using FEA to initialize the surrogate. The surrogate is then used for the microscale solutions in the following computations when there are enough neighboring points to perform the approximation. $I\left(\bar{\varepsilon}^*, \bar{\varepsilon}(t_k)\right)$ in algorithm 1.2 is a counting function

$$I\left(\bar{\varepsilon}^*, \bar{\varepsilon}(t_k)\right) = \begin{cases} 1, \text{ if } \operatorname{dist}\left(\bar{\varepsilon}^*, \bar{\varepsilon}(t_k)\right) \leq r_{\text{diff}}, \\ 0, \text{ if } \operatorname{dist}\left(\bar{\varepsilon}^*, \bar{\varepsilon}(t_k)\right) > r_{\text{diff}}, \end{cases} \quad [2.22]$$

which counts the number of points within the local influence zone r_{diff}

$$r_{\text{diff}} = \frac{\text{dist}\left((\bar{\varepsilon}_{11}^{\max}, \bar{\varepsilon}_{22}^{\max}, \bar{\varepsilon}_{12}^{\max}), (\bar{\varepsilon}_{11}^{\min}, \bar{\varepsilon}_{22}^{\min}, \bar{\varepsilon}_{12}^{\min})\right)}{N_{\text{ratio}}}, \qquad [2.23]$$

defined as a ratio N_{ratio} of the Euclidean distance between the maximum and minimum strain components in the surrogate. N_{approx} is the required number that is given in accordance with the surrogate size and also the order of the applied polynomial.

Algorithm 2.1. Microscale RVE solutions

1: given $\bar{\varepsilon}^*, i, j, k$;
2: define PBC on RVE upon $\bar{\varepsilon}^*$;
3: **if** $i = 1$ **and** $j = 1$ **then**
4:     solve the RVE problem using FEA;
5:     extract $\mathbf{\Phi}$ and saving the coefficients;
6: **else if** $\sum_k I\left(\bar{\varepsilon}^*, \bar{\varepsilon}(t_k)\right) < N_{\text{approx}}$ **or** $k \geq N_{\text{sub}}$ **then**
7:     solve the RVE problem using FEA;
8:     update $\mathbf{\Phi}$ and enrich the surrogate database;
9: **else**
10:     solve the RVE problem using the surrogate;
11: **end if**
12: compute the effective stress $\bar{\boldsymbol{\sigma}}^*$;
13: compute the effective tangent stiffness $\bar{\mathbf{C}}^{\tan}$;
14: **return** $\bar{\boldsymbol{\sigma}}^*$ and $\bar{\mathbf{C}}^{\tan}$.

When there is not enough points within the local influence zone, the microproblem is solved using full FEA and the results are used to update the POD basis $\mathbf{\Phi}$ and enrich the surrogate database. In this work, the scale of the surrogate is kept constant after its initialization, which means the enrichment includes current analysis results while excluding the same number of previous results. On the one hand, the previous results no longer contribute in the following designs as the structural topology varies; on the other hand, the Diffuse Approximation becomes more and more expensive in computing as the scale of the surrogate database grows.

In the case where the macroscopic convergence in algorithm 1.1 cannot be reached after a certain number of iterations, i.e. $||\bar{\mathbf{f}}_{\text{ext}} - \bar{\mathbf{f}}_{\text{int}}|| > \delta_{\text{f}}$, we can either shrink the local influence zone r_{diff}, or force it to use full FEA to update the surrogate. The later solution is recommended in the consideration of both aspects of analysis accuracy and the convergence efficiency. In this algorithm, full FEA is applied to perform microscale computation after a certain number of macroscopic iterations, i.e. $k \geq N_{\text{sub}}$.

2.3. Numerical examples

The benchmark cantilever problem [ALL 04b] is considered in this section with anisotropic material defined at the microscopic scale. As illustrated in Figure 2.3, the macroscopic structure is discretized into 32×20 four-node plane strain elements where each element has four Gauss integration points. Each Gauss point at the macroscopic structure corresponds to a considered RVE at the microscopic scale. The material property of the solid phase in the RVE is assumed to be isotropic with a nonlinear elastic constitutive behavior as shown in Figure 2.3. Conventional unreduced FE^2 approach requires $32 \times 20 \times 4$ independent RVE analyses at the microscopic scale for one time evaluation of the macroscopic equilibrium. For the sake of simplicity, the initial elastic stiffness matrix has been kept during the NR iterative solution procedure. In order to perform sensitivity analysis, tangent stiffness matrix is evaluated using the perturbation method only at the convergence of the solution.

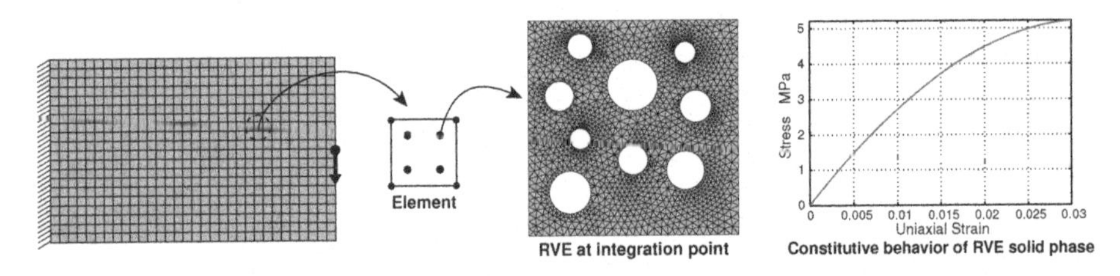

Figure 2.3. *Illustration of a two-scale cantilever made of nonlinear porous material*

2.3.1. *Test case 1*

In the first test case, the value of the external loading force is set to 0.5 N and the considered volume constraint is 40%. The parameters in [2.7] are set as $\alpha = 0.9, \lambda = -10^{-8}, \Lambda = 4 \times 10^{7}$. The tolerance error in [2.15] is set as

$\delta = 10^{-6}$. The extracted POD modes vary adaptively during the optimization procedure and the size of the reduced basis is 5 and stays constant. In Figure 2.4, we show the resultant tractions used in [1.9] of the first five final POD modes along with the associated normalized eigenvalues $\lambda_i/\lambda_{\max}$. The ratio in [2.23] to define the influence zone is set as $N_{\text{ratio}} = 20$, and the required number of approximating points in algorithm 2.1 is set as $N_{\text{approx}} = 7$.

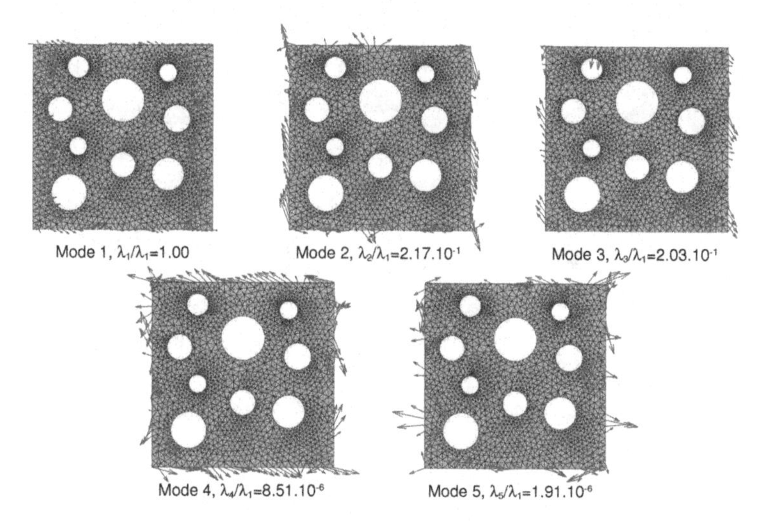

Figure 2.4. *Resultant tractions on the RVE boundaries of the first five POD modes. For a color version of the figure, see www.iste.co.uk/xia/topology.zip*

The structural topological evolution at the macroscopic scale is given in Figure 2.5. Usually, topology optimization gives a symmetric design result when isotropic material constitutive behavior is considered due to the symmetry of the cantilever problem. However, the final structural design obtained here is no longer symmetric due to the anisotropy of the considered RVE at the microscopic scale. The convergence histories of the strain energy and the volume ratio are demonstrated in Figures 2.6(a) and (b), respectively.

During the loading phase of the first optimization iteration, the periodic homogenizations of the RVE at the microscopic scale are performed using full FEA. Since the second optimization iteration, both FEA and the surrogate model are used for the microscopic analysis. Figure 2.6(c) gives the percentage of FEA usage in each optimization iteration. It can be seen that less than 4% microscopic analysis require full FEA due to the usage of a surrogate model, which significantly reduces computational cost. The jumps in Figure 2.6(c) correspond to comparatively large volume and topological variations, which require more usage of FEA in order to update the POD basis and enrich the surrogate database.

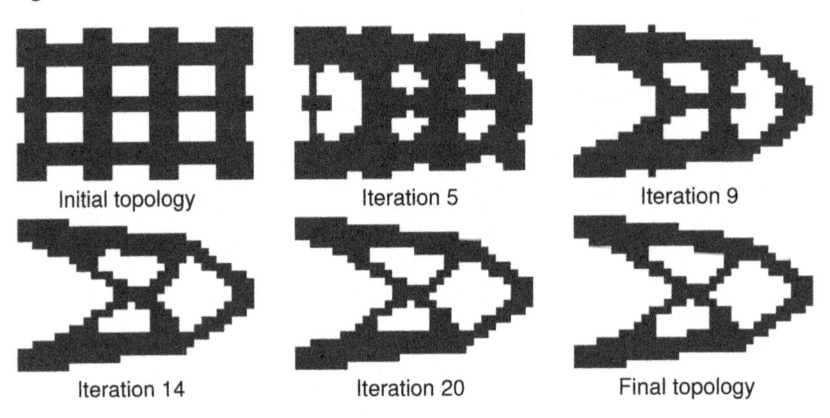

Figure 2.5. *Topological evolution of the design process*

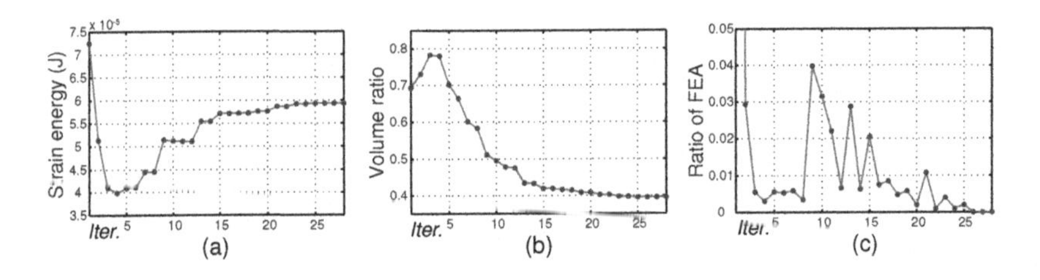

Figure 2.6. *Optimization history: a) convergence history of the strain energy, b) convergence history of the volume ratio and c) the ratio of FEA usage in each iteration*

The same optimization design has also been performed without using the surrogate. The unreduced FE^2 approach gives exactly the same optimization design result as the reduced model where the relative errors of the objective are less than 10^{-5}. The iterative computing time using the unreduced FE^2 decreases as the volume ratio decreases (see Figure 2.6(b), i.e. the number of microanalyses required in each substep of macroscopic computing decreases from the maximum $0.8 \times 32 \times 20 \times 4$ to finally $0.4 \times 32 \times 20 \times 4$. Generally speaking, it requires around 2 h of computing for each optimization iteration on a HP Z420 Workstation when using the unreduced sequential FE^2. In contrast, the reduced FE^2 approach requires only 10 min of computing on average for each design iteration apart from the first design iteration. More saving in computation can be expected using the reduced approach when larger scale problems are considered.

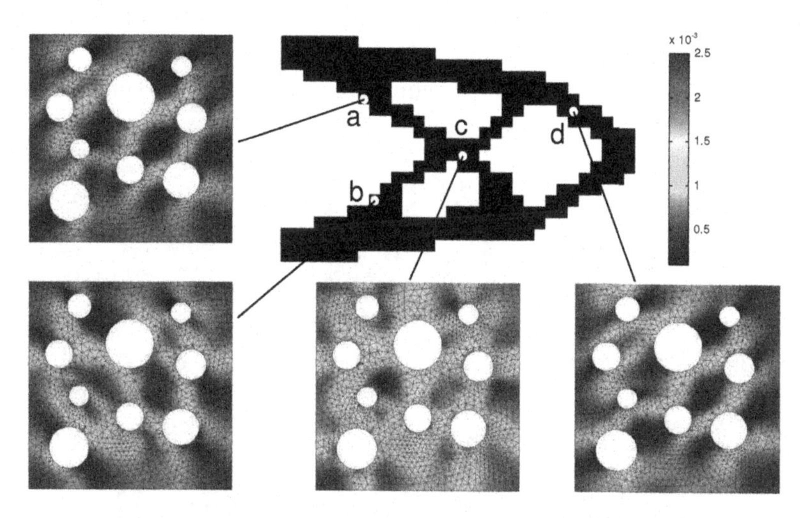

Figure 2.7. *Microscale equivalent strain distributions at selected points. For a color version of the figure, see www.iste.co.uk/xia/topology.zip*

Figure 2.7 depicts the equivalent strain distributions at the microscopic scale at selected points. We may note that the existence of the holes in the RVE concentrates much higher strains and hence stresses at the microscopic scale than the homogenized macroscopic values. Moreover, different microstrain distributions manifest the difference of the local loading statuses of the selected points, where point c in the center of the structure obviously suffers more mechanical loads than typical points (a,b,d) located off the main loading region. The higher stress concentration may result in the initial material failure or crack at the microscopic scale.

2.3.2. *Test case 2*

In order to evaluate the performance of the surrogate model when encountering more nonlinearity and more severe topological changes, the external loading force is increased to 1.5 N and the considered volume ratio constraint is decreased to 32%. The parameters in [2.7] are set as $\alpha = 0.9, \lambda = -10^{-7}, \Lambda = 7 \times 10^{6}$. The tolerance error in [2.15] is set as in the previous case of $\delta = 10^{-6}$. Correspondingly, the number of retained POD modes increases to six during the first iterations and then to seven during the following iterations until the end. The resultant tractions of the first seven of the final POD modes are shown in Figure 2.8 together with their associated normalized eigenvalues $\lambda_i/\lambda_{\max}$. The ratio in [2.23] to define the influence zone is set as $N_{\text{ratio}} = 20$, and the required number of approximating points in algorithm 21 is set as $N_{\text{approx}} = 7$.

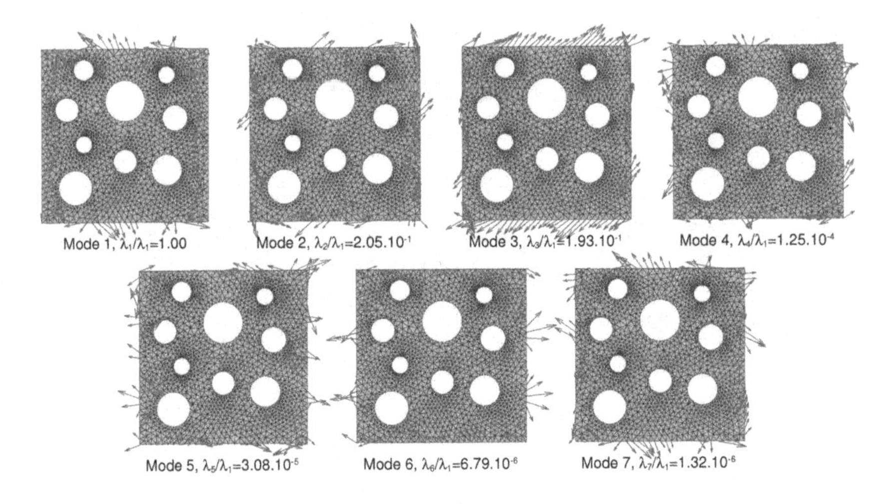

Figure 2.8. *Resultant tractions on the RVE boundaries of the first seven POD modes. For a color version of the figure, see www.iste.co.uk/xia/topology.zip*

The structural topological evolution at the macroscopic scale given in Figure 2.9 is similar to the previous case during the first iterations while it differs later due the applied lower volume ratio constraint. The convergence histories of the strain energy and the volume ratio are demonstrated in Figures 2.10(a) and (b), respectively.

Likewise, the surrogate model is initialized during the loading phase of the first optimization iteration. Both the surrogate model and FEA are used in the following optimization iterations. Figure 2.10(c) gives the percentage of FEA usage in each optimization iteration. Similar to the previous case, less than

4% microscopic analysis requires full FEA except a jump from 2% in iteration 20 to 17% in iteration 21. A detailed illustration of the topological evolution around iteration 21 is given in Figure 2.11. It can be seen that a branch of the structure splits in iteration 21. Such a severe topological variation results in a large variation of the structural physical response and hence the surrogate built according to the previous calculations is no longer accurate enough. Therefore, an increased number of full FEA is required to recompute the set of reduced basis. The surrogate model is updated thereafter and the usage ratio of FEA drops back below 4% and decreases to 0% in the following iterations as the structural topology converges, meaning that all computations are performed using the surrogate model.

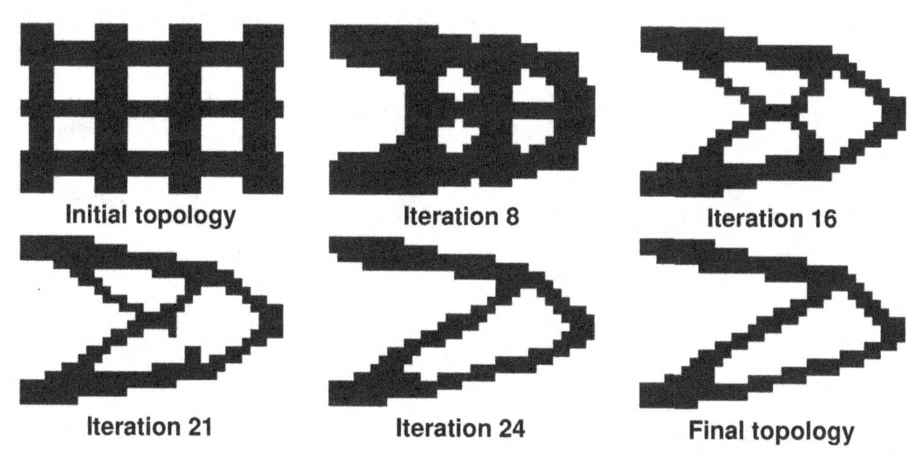

Figure 2.9. *Topological evolution of the design process*

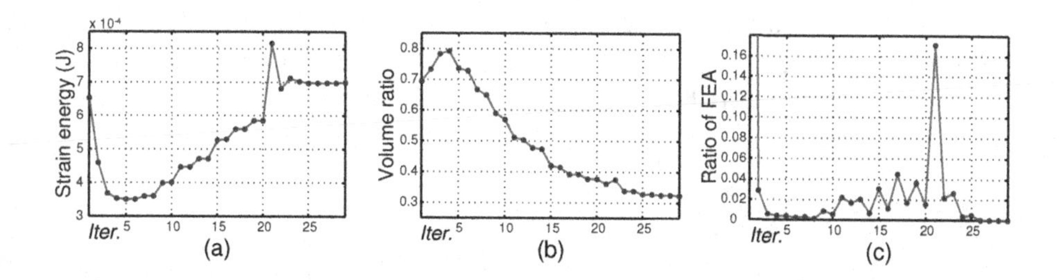

Figure 2.10. *Optimization history: a) convergence history of the strain energy, b) convergence history of the volume ratio and c) the ratio of FEA usage in each iteration*

For the purpose of comparison, the same optimization design has also been performed using the unreduced FE^2 approach that gives exactly the same optimization design result as obtained above where the relative errors of the objective for each design iteration are less than 10^{-5}. As shown in Figure 2.10(b), the number of microanalyses required in each substep of macroscopic computing varies from the maximum $0.8 \times 32 \times 20 \times 4$ to finally $0.32 \times 32 \times 20 \times 4$. Due to the increased nonlinearity, more substeps have to be taken to reach the macroconvergence and thereafter the average computing time for each optimization augments to around 3 h. Similarly, more substeps are required to reach the macroconvergence when using the surrogate, and average computing time required by the reduced approach increases to 15 min, which is nevertheless a significant gain in time compared to the time required using the unreduced approach.

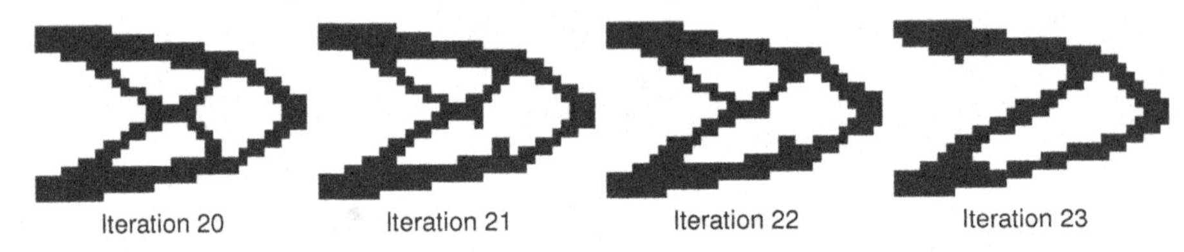

Figure 2.11. *Intermediate design topologies from iteration* 20 *to iteration* 23

Figure 2.12 depicts the equivalent strain distributions at the microscopic scale at selected points. Because of the increase in the external loading force and the decrease in the volume ratio constraint, larger deformations can be observed from the equivalent strain distributions at the microscopic scale. The microstrain distributions clearly manifest the difference of the loading status in different structural branches. The microstrain distributions at points b and c are quite similar because they are located in the same branch of the structure.

2.4. Concluding remarks

We have developed in this chapter a POD-based adaptive surrogate model for the solutions at the microscopic scale dedicated to reduce the heavy computational burden for the multiscale design framework developed in Chapter 1. The surrogate model has shown promising performance in terms of reducing computing cost and modeling accuracy when applied to the design framework for nonlinear elastic cases.

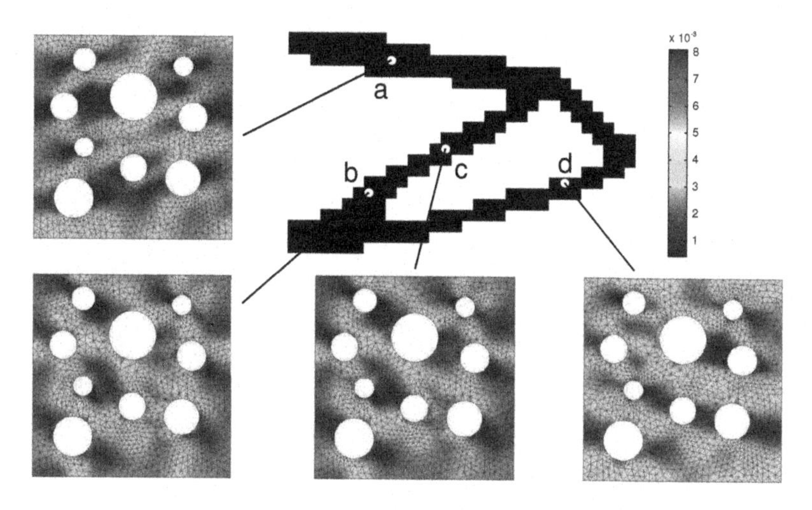

Figure 2.12. *Microscale equivalent strain distribution at selected points. For a color version of the figure, see www.iste.co.uk/xia/topology.zip*

For moderate nonlinearities such as the case considered here, it could be an easier solution to build the interpolating relationship directly between the effective stress and strain that may perform well. However, the present strategy using the field interpolation is more accurate and the on-line learning procedure is more suitable for capturing nonlinear features.

In the following chapter, we will take a step further toward the design of multiscale elastoviscoplastic structures. In order to realize such design in realistic computing times and with affordable memory requirements, we will directly employ the pRBMOR method, a specifically developed ROM for material nonlinear homogenization involving viscoplasticities. In addition, the computing time will be drastically reduced by implementing GPU parallelization.

3

Topology Optimization of Multiscale Elastoviscoplastic Structures

In this chapter, we take a step further toward the design of multiscale elastoviscoplastic structures (Figure 3.1) using the multiscale design framework discussed in Chapter 1. This subject is extremely challenging from both aspects of *topology optimization* and *multiscale modeling*. First, unlike linear designs, topology optimization of elastovisoplastic structures encounters instability issues during the iterative solution process and the evaluation of sensitivities is more demanding. Second, the consideration of path-dependent plastic behavior at the microscopic scale results in significantly augmented computational burden in terms of computing time and storage requirement when using the FE^2 method.

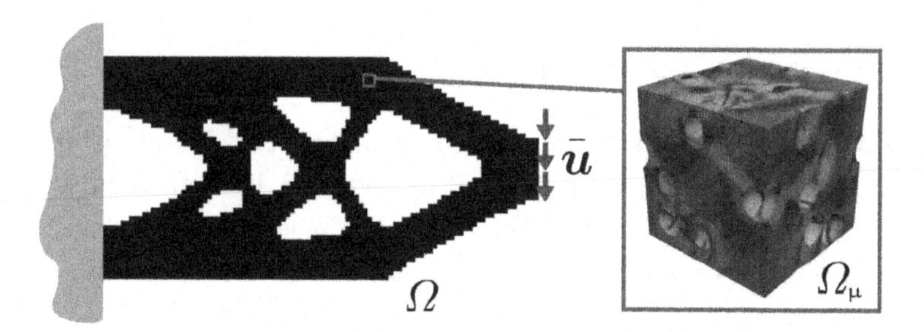

Figure 3.1. *Design of a two-scale structure made up of porous elastoviscoplastic materials [FRI 15b]. For a color version of the figure, see www.iste.co.uk/xia/topology.zip*

With regard to topology optimization, the BESO method is applied again for its algorithmic advantage of the discrete nature in designing multiscale structures. Unlike the previous versions of the BESO updating scheme for linear or nonlinear elastic designs, a stabilization scheme controlling the number of recovered elements is implemented in order to stabilize the design process. In addition, the sensitivities of the design variables for nonlinear dissipative problems are derived in a clear and rigorous manner using the adjoint method.

In order to alleviate the computational burden of FE^2, the computationally demanding nonlinear microscale problem is substituted by the pRBMOR method [FRI 13b, FRI 14, FRI 15a]. This part of work was performed in collaboration with Felix Fritzen from the University of Stuttgart [FRI 15b]. Our contribution consists of the modification of the proposed multiscale design framework to accommodate the pRBMOR method. In addition, the pRBMOR method is implemented in parallel on GPUs [FRI 14, FRI 15a], which further speeds up the computation.

This chapter is organized in the following manner: the multiscale design framework is reviewed in section 3.1 with a detailed derivation of the sensitivity analysis and an implementation of a stabilization scheme controlling material addition. Section 3.2 presents the microscopic modeling of an elastoviscoplastic porous material and a brief summary of pRBMOR. In section 3.3, numerical examples of a viscoplastic porous metal with nonlinear hardening are investigated. Concluding remarks are given in section 3.4.

3.1. Topology optimization model

The following framework is presented with the consideration of the multiscale problem setting (FE^2) presented in section 1.1. Matrix and vector forms are applied in accordance with the FEM. The macroscopic stress $\bar{\sigma}$ and strain $\bar{\varepsilon}$ represent their vector forms following the Voigt notation [1.10].

3.1.1. *Model definitions*

The basic definitions are in analogy to the definitions given in section 1.4.1. We have the design domain Ω discretized into N_{e} finite elements and each element e is assigned a topology design variable ρ_e. Topology design variables are defined in analogy to [1.16] that the macroscopic element internal force vector $\bar{\mathbf{f}}^e_{\text{int}}$ is associated topology variable ρ_e

$$\bar{\mathbf{f}}^e_{\text{int}} = \rho_e \int_{\Omega_e} \bar{\mathbf{B}}^T \bar{\boldsymbol{\sigma}} \mathrm{d}\Omega_e. \qquad [3.1]$$

The effective stress $\bar{\boldsymbol{\sigma}}$ occurring in [3.1] is computed via the volume averaging relation $\bar{\boldsymbol{\sigma}} = \langle \boldsymbol{\sigma} \rangle$, in which the microscopic stress $\boldsymbol{\sigma}$ is determined from an underlying nonlinear microscale simulation subjected to a prescribed overall strain $\bar{\boldsymbol{\varepsilon}}$. More precisely, not only is the current load required, but due to the dissipative material behavior, the entire load path has to be applied to the RVE in order to account for the path dependency of the microscopic (and thus macroscopic) constitutive response. In practice, for void elements the effective stress is set to zero.

Displacement-controlled loading is applied in the sequel due to stability considerations (e.g. [MAU 98, SCH 01, CHO 03, YOO 07, HUA 07b, HUA 08]). In the following, the objective of the design optimization is to maximize the structural stiffness. This is equivalent to the maximization of the mechanical work expended in the course of the deformation process. In practice, the total mechanical work, which is expressed in an equivalent form of the complementary external work, f_{w} is approximated by numerical integration using the trapezoidal rule, i.e.

$$f_{\text{w}} \approx \frac{1}{2} \sum_{i=1}^{n_{\text{load}}} \left(\bar{\mathbf{f}}^{(i)}_{\text{ext}} + \bar{\mathbf{f}}^{(i-1)}_{\text{ext}} \right)^T \Delta \bar{\mathbf{u}}^{(i)}. \qquad [3.2]$$

Here, n_{load} is the total number of displacement increments, $\Delta \bar{\mathbf{u}}^{(i)}$ is the ith increment of the nodal displacement vector and $\bar{\mathbf{f}}^{(i)}_{\text{ext}}$ is the external nodal force vector at the ith load increment. During the optimization, the material volume fraction is prescribed. Then, the optimization problem can be formulated as [HUA 08]

$$\begin{aligned} \max_{\boldsymbol{\rho}} &: f_{\text{w}}(\boldsymbol{\rho}, \bar{\mathbf{u}}) \\ \text{subject to} &: \bar{\mathbf{r}}(\boldsymbol{\rho}, \bar{\mathbf{u}}) = \mathbf{0} \\ &: V(\boldsymbol{\rho}) = \sum \rho_e v_e = V_{\text{req}} \\ &: \rho_e = 0 \text{ or } 1, e = 1, \ldots, N_{\text{e}}. \end{aligned} \qquad [3.3]$$

Here, $V(\boldsymbol{\rho})$ is the total volume of solid elements, V_{req} is the required volume of solid elements and v_e is the volume of element e. $\bar{\mathbf{u}}$ is the displacement solution at convergence and $\bar{\mathbf{r}}(\boldsymbol{\rho}, \bar{\mathbf{u}})$ is the residual at the macroscopic scale defined in analogy to [1.20]

$$\bar{\mathbf{r}} = \bar{\mathbf{f}}_{\text{ext}} - \sum_{e=1}^{N_e} \rho_e \int_{\Omega_e} \bar{\mathbf{B}}^T \bar{\boldsymbol{\sigma}} \mathrm{d}\Omega_e. \qquad [3.4]$$

3.1.2. *Sensitivity analysis*

In order to perform topology optimization, the sensitivity of the objective function f_{w} with respect to the topology design variables $\boldsymbol{\rho}$ needs to be provided. Similar to the procedure presented in section 1.4.2, the derivation of the sensitivity requires using the adjoint method (see [BUH 00, CHO 03]). Lagrangian multipliers $\boldsymbol{\mu}^{(i)}$, $\boldsymbol{\lambda}^{(i)}$ of the same dimension as the vector of unknowns $\bar{\mathbf{u}}$ are introduced in order to enforce zero residual $\bar{\mathbf{r}}$ at times t_{i-1} and t_i for each term of the quadrature rule [3.2]. Then, the objective function f_{w} can be rewritten in the following form without modifying the original objective value as

$$f_{\text{w}}^* = \frac{1}{2} \sum_{i=1}^{n_{\text{load}}} \left(\bar{\mathbf{f}}_{\text{ext}}^{(i)} + \bar{\mathbf{f}}_{\text{ext}}^{(i-1)} \right)^T \Delta \bar{\mathbf{u}}^{(i)} + \left(\boldsymbol{\lambda}^{(i)} \right)^T \bar{\mathbf{r}}^{(i)} + \left(\boldsymbol{\mu}^{(i)} \right)^T \bar{\mathbf{r}}^{(i-1)}. \qquad [3.5]$$

Due to the asserted static equilibrium, the residuals $\bar{\mathbf{r}}^{(i)}$ and $\bar{\mathbf{r}}^{(i-1)}$ have to vanish. The objective value is, thus, invariant with respect to the values of the Lagrangian multipliers $\boldsymbol{\lambda}^{(i)}$ and $\boldsymbol{\mu}^{(i)}$ ($i = 1, \ldots, n_{\text{load}}$), i.e.

$$f_{\text{w}}^* \left(\boldsymbol{\rho}; \left\{ \boldsymbol{\lambda}^{(i)}, \boldsymbol{\mu}^{(i)} \right\}_{i=1,\ldots,n_{\text{load}}} \right) = f_{\text{w}} (\boldsymbol{\rho}). \qquad [3.6]$$

This equivalence also holds for the sensitivity with respect to changes in the order parameter ρ_e in the element e

$$\frac{\partial f_{\text{w}}^*}{\partial \rho_e} = \frac{\partial f_{\text{w}}}{\partial \rho_e}. \qquad [3.7]$$

In the following, $\partial f_{\text{w}}^* / \partial \rho_e$ is computed with properly determined values of $\boldsymbol{\lambda}^{(i)}$ and $\boldsymbol{\mu}^{(i)}$ leading to certain simplifications of the derivation. In order to formally describe these derivations, we introduce a partitioning of all degrees of freedom (DOF) into essential (index E; associated with Dirichlet boundary conditions) and free (index F; remaining DOF) entries. For a vector $\mathbf{w}$ and a matrix $\mathbf{M}$, we have

$$\mathbf{w} \sim \begin{bmatrix} \mathbf{w}_{\mathrm{E}} \\ \mathbf{w}_{\mathrm{F}} \end{bmatrix}, \qquad \mathbf{M} \sim \begin{bmatrix} \mathbf{M}_{\mathrm{EE}} & \mathbf{M}_{\mathrm{EF}} \\ \mathbf{M}_{\mathrm{FE}} & \mathbf{M}_{\mathrm{FF}} \end{bmatrix}. \qquad [3.8]$$

In the current context, the displacements $\bar{\mathbf{u}}_{\mathrm{E}}$ on the Dirichlet boundary are prescribed and, hence, they are independent of the current value of $\boldsymbol{\rho}$. This implies that

$$\frac{\partial \Delta \bar{\mathbf{u}}}{\partial \rho_e} = \frac{\partial}{\partial \rho_e} \begin{bmatrix} \Delta \bar{\mathbf{u}}_{\mathrm{E}} \\ \Delta \bar{\mathbf{u}}_{\mathrm{F}} \end{bmatrix} = \begin{bmatrix} \mathbf{0} \\ \partial \left(\Delta \bar{\mathbf{u}}_{\mathrm{F}} \right) / \partial \rho_e \end{bmatrix} \qquad [3.9]$$

holds for arbitrary times t, i.e. for $\bar{\mathbf{u}} = \bar{\mathbf{u}}^{(i)}$ or $\bar{\mathbf{u}} = \bar{\mathbf{u}}^{(i-1)}$. The components $\bar{\mathbf{f}}_{\mathrm{ext,F}}$ of the force vector $\bar{\mathbf{f}}_{\mathrm{ext}}$ vanish at all times t_i and the only (possibly) non-zero components are the reaction forces $\bar{\mathbf{f}}_{\mathrm{ext,E}}$

$$\bar{\mathbf{f}}_{\mathrm{ext}}^{(i)} = \begin{bmatrix} \bar{\mathbf{f}}_{\mathrm{ext,E}}^{(i)} \\ \mathbf{0} \end{bmatrix}. \qquad [3.10]$$

equations [3.9] and [3.10] imply

$$\left(\bar{\mathbf{f}}_{\mathrm{ext}}^{(j)} \right)^T \frac{\partial \Delta \bar{\mathbf{u}}^{(i)}}{\partial \rho_e} = 0. \qquad [3.11]$$

hence, the expansion

$$\frac{\partial}{\partial \rho_e} \left(\left(\bar{\mathbf{f}}_{\mathrm{ext}}^{(j)} \right)^T \Delta \bar{\mathbf{u}}^{(i)} \right) = \left(\frac{\partial \bar{\mathbf{f}}_{\mathrm{ext}}^{(j)}}{\partial \rho_e} \right)^T \Delta \bar{\mathbf{u}}^{(i)} + \left(\bar{\mathbf{f}}_{\mathrm{ext}}^{(j)} \right)^T \frac{\partial \Delta \bar{\mathbf{u}}^{(i)}}{\partial \rho_e} \qquad [3.12]$$

for arbitrary time indices $i, j = 1, \ldots, n_{\mathrm{load}}$ can be reduced into

$$\frac{\partial}{\partial \rho_e} \left(\left(\bar{\mathbf{f}}_{\mathrm{ext}}^{(j)} \right)^T \Delta \bar{\mathbf{u}}^{(i)} \right) = \left(\frac{\partial \bar{\mathbf{f}}_{\mathrm{ext}}^{(j)}}{\partial \rho_e} \right)^T \Delta \bar{\mathbf{u}}^{(i)}. \qquad [3.13]$$

therefore, the sensitivity of the modified objective in [3.5] equals

$$\frac{\partial f_{\mathrm{w}}^*}{\partial \rho_e} = \frac{1}{2} \sum_{i=1}^{n_{\mathrm{load}}} \left[\left(\frac{\partial \bar{\mathbf{f}}_{\mathrm{ext}}^{(i)}}{\partial \rho_e} + \frac{\partial \bar{\mathbf{f}}_{\mathrm{ext}}^{(i-1)}}{\partial \rho_e} \right)^T \Delta \bar{\mathbf{u}}^{(i)} + \left(\boldsymbol{\lambda}^{(i)} \right)^T \frac{\partial \bar{\mathbf{r}}^i}{\partial \rho_e} + \left(\boldsymbol{\mu}^{(i)} \right)^T \frac{\partial \bar{\mathbf{r}}^{i-1}}{\partial \rho_e} \right]. \qquad [3.14]$$

According to the residual definition [3.4], the derivatives of $\bar{\mathbf{r}}^{(j)}$ at the equilibrium of the jth load increment with respect to ρ_e can be expanded as

$$\frac{\partial \bar{\mathbf{r}}^{(j)}}{\partial \rho_e} = \frac{\partial \bar{\mathbf{f}}_{\text{ext}}^{(j)}}{\partial \rho_e} - \int_{\Omega_e} \bar{\mathbf{B}}^T \bar{\boldsymbol{\sigma}}^{(j)} \mathrm{d}\Omega_e - \bar{\mathbf{K}}_{\tan}^{(j)} \frac{\partial \Delta \bar{\mathbf{u}}^{(j)}}{\partial \rho_e}, \quad [3.15]$$

with

$$\bar{\mathbf{K}}_{\tan}^{(j)} = -\frac{\partial \bar{\mathbf{r}}^{(j)}}{\partial \bar{\mathbf{u}}^{(j)}} \quad [3.16]$$

the global FE tangent stiffness matrix of the nonlinear mechanical system at the equilibrium of the jth load increment. With the result of [3.15], [3.14] can be reformulated as

$$\begin{aligned} \frac{\partial f_{\text{w}}^*}{\partial \rho_e} = \frac{1}{2} \sum_{i=1}^{n_{\text{load}}} \Bigg[& \left(\frac{\partial \bar{\mathbf{f}}_{\text{ext}}^{(i)}}{\partial \rho_e} \right)^T \left(\Delta \bar{\mathbf{u}}^{(i)} + \boldsymbol{\lambda}^{(i)} \right) + \left(\frac{\partial \bar{\mathbf{f}}_{\text{ext}}^{(i-1)}}{\partial \rho_e} \right)^T \left(\Delta \bar{\mathbf{u}}^{(i)} + \boldsymbol{\mu}^{(i)} \right) \\ & - \left(\boldsymbol{\lambda}^{(i)} \right)^T \int_{\Omega_e} \bar{\mathbf{B}}^T \bar{\boldsymbol{\sigma}}^{(i)} \mathrm{d}\Omega_e - \left(\boldsymbol{\mu}^{(i)} \right)^T \int_{\Omega_e} \bar{\mathbf{B}}^T \bar{\boldsymbol{\sigma}}^{(i-1)} \mathrm{d}\Omega_e \\ & - \left(\boldsymbol{\lambda}^{(i)} \right)^T \bar{\mathbf{K}}_{\tan}^{(i)} \frac{\partial \Delta \bar{\mathbf{u}}^{(i)}}{\partial \rho_e} - \left(\boldsymbol{\mu}^{(i)} \right)^T \bar{\mathbf{K}}_{\tan}^{(i-1)} \frac{\partial \Delta \bar{\mathbf{u}}^{(i-1)}}{\partial \rho_e} \Bigg]. \end{aligned} \quad [3.17]$$

As mentioned before, the aim is to find proper values of the Lagrangian multipliers $\boldsymbol{\lambda}^{(i)}$ and $\boldsymbol{\mu}^{(i)}$ such that the sensitivities can be explicitly and efficiently computed. From the consideration of [3.10], the first two terms can be canceled by setting

$$\boldsymbol{\lambda}_{\text{E}}^{(i)} = -\Delta \bar{\mathbf{u}}_{\text{E}}^{(i)} \quad \text{and} \quad \boldsymbol{\mu}_{\text{E}}^{(i)} = -\Delta \bar{\mathbf{u}}_{\text{E}}^{(i)}. \quad [3.18]$$

Accounting further for the structure of the sensitivities of $\bar{\mathbf{u}}$ in [3.9] and for the symmetry of the tangent stiffness operator, we have

$$\begin{aligned} \frac{\partial f_{\text{w}}^*}{\partial \rho_e} = \frac{1}{2} \sum_{i=1}^{n_{\text{load}}} \Bigg[& - \left(\boldsymbol{\lambda}^{(i)} \right)^T \int_{\Omega_e} \bar{\mathbf{B}}^T \bar{\boldsymbol{\sigma}}^{(i)} \mathrm{d}\Omega_e - \left(\boldsymbol{\mu}^{(i)} \right)^T \int_{\Omega_e} \bar{\mathbf{B}}^T \bar{\boldsymbol{\sigma}}^{(i-1)} \mathrm{d}\Omega_e \\ & - \left(\bar{\mathbf{K}}_{\tan,\text{FE}}^{(i)} \boldsymbol{\lambda}_{\text{E}}^{(i)} + \bar{\mathbf{K}}_{\tan,\text{FF}}^{(i)} \boldsymbol{\lambda}_{\text{F}}^{(i)} \right)^T \frac{\partial \Delta \bar{\mathbf{u}}_{\text{F}}^{(i)}}{\partial \rho_e} \\ & - \left(\bar{\mathbf{K}}_{\tan,\text{FE}}^{(i-1)} \boldsymbol{\mu}_{E}^{(i)} + \bar{\mathbf{K}}_{\tan,\text{FF}}^{(i-1)} \boldsymbol{\mu}_{F}^{(i)} \right)^T \frac{\partial \Delta \bar{\mathbf{u}}_{\text{F}}^{(i-1)}}{\partial \rho_e} \Bigg]. \end{aligned} \quad [3.19]$$

In order to avoid evaluating the unknown derivatives of $\bar{\mathbf{u}}_{\mathrm{F}}^{(i)}$ and $\bar{\mathbf{u}}_{\mathrm{F}}^{(i-1)}$, i.e. eliminating the last two lines of [3.19], the values of $\boldsymbol{\lambda}_{\mathrm{F}}^{(i)}$ and $\boldsymbol{\mu}_{\mathrm{F}}^{(i)}$ are sought as following by solving the adjoint systems with the prescribed values $\boldsymbol{\lambda}_{\mathrm{E}}^{(i)} = -\Delta\bar{\mathbf{u}}_{\mathrm{E}}^{(i)}$ and $\boldsymbol{\mu}_{\mathrm{E}}^{(i)} = -\Delta\bar{\mathbf{u}}_{\mathrm{E}}^{(i)}$ at the essential nodes:

$$\boldsymbol{\lambda}_{\mathrm{F}}^{(i)} = \left(\bar{\mathbf{K}}_{\mathrm{tan,FF}}^{(i)}\right)^{-1} \bar{\mathbf{K}}_{\mathrm{tan,FE}}^{(i)} \Delta\bar{\mathbf{u}}_{\mathrm{E}}^{(i)}, \qquad [3.20]$$

and

$$\boldsymbol{\mu}_{\mathrm{F}}^{(i)} = \left(\bar{\mathbf{K}}_{\mathrm{tan,FF}}^{(i-1)}\right)^{-1} \bar{\mathbf{K}}_{\mathrm{tan,FE}}^{(i-1)} \Delta\bar{\mathbf{u}}_{\mathrm{E}}^{(i)}. \qquad [3.21]$$

These two relations [3.20] and [3.21] together with [3.18] completely determine the values of Lagrangian multipliers $\boldsymbol{\lambda}^{(i)}$ and $\boldsymbol{\mu}^{(i)}$. Finally, $\partial f_{\mathrm{w}}^{*}/\partial \rho_e$ can be computed via

$$\frac{\partial f_{\mathrm{w}}^{*}}{\partial \rho_e} = -\frac{1}{2} \sum_{i=1}^{n_{\mathrm{load}}} \left[\left(\boldsymbol{\lambda}^{(i)}\right)^T \int_{\Omega_e} \bar{\mathbf{B}}^T \bar{\boldsymbol{\sigma}}^{(i)} \mathrm{d}\Omega_e + \left(\boldsymbol{\mu}^{(i)}\right)^T \int_{\Omega_e} \bar{\mathbf{B}}^T \bar{\boldsymbol{\sigma}}^{(i-1)} \mathrm{d}\Omega_e \right]. \qquad [3.22]$$

The computation of the sensitivity persists in solving two linear systems of equations once for all elements. Note that because the proportional loading is increased at a constant rate $\dot{\bar{\mathbf{u}}}_{\mathrm{E}}^{0}$, i.e.

$$\Delta\bar{\mathbf{u}}_{\mathrm{E}}^{(i)} = \Delta t^{(i)} \dot{\bar{\mathbf{u}}}_{\mathrm{E}}^{0}, \qquad [3.23]$$

the solution of the second linear system can therefore be omitted by means of the recursion formula

$$\boldsymbol{\mu}_{\mathrm{F}}^{(i)} = \frac{\Delta t^{(i)}}{\Delta t^{(i-1)}} \boldsymbol{\lambda}_{\mathrm{F}}^{(i-1)}. \qquad [3.24]$$

3.1.3. *BESO updating scheme*

The BESO updating scheme for the design of elastoviscoplastic structures follows in great part the basic procedure presented in section 1.4.3, except for the addition of a stabilization scheme controlling the number of recovered elements at each design iteration and minor modifications on certain implementations.

In contrast to the proposal applied in section 1.4.3 for the determination of the volume of material usage at the current design iteration, a linear decay of the volume fraction is used in the present work, i.e.

$$V^{(l)} = \max\left\{V_{\text{req}}, V^{(l-1)} - c_{\text{er}}V^{(0)}\right\}. \qquad [3.25]$$

Here, the evolutionary ratio c_{er} determines the percentage of material to be removed and $V^{(0)}$ is the initial material volume corresponding to full solid structure in the current context.

The sensitivity numbers of the elements that are used to determine material removal and addition is defined as

$$\alpha_e = \frac{\partial f_{\text{w}}^*}{\partial \rho_e}\frac{1}{v_e}, \qquad [3.26]$$

The sensitivity numbers are modified further to remove numerical artifacts by the linear filtering scheme [1.29]. The filter is also responsible for material addition as the sensitivity in a void element is zero. An interpretation of the filter is that void regions next to highly loaded regions are considered as highly sensitive to the overall response as switching them back to solid elements can lead to an unloading in the highly loaded surrounding solid elements.

Note that at certain design iterations when severe topological changes occur, the sensitivity numbers in the affected zones are several orders of magnitude larger than the sensitivity numbers at other design iterations; the introduction of this abnormal historical information as in [1.31]

$$\alpha_e^{(l)} \leftarrow (\alpha_e^{(l)} + \alpha_e^{(l-1)})/2 \qquad [3.27]$$

will no longer stabilize but perturb the following design evolution. Therefore, unlike in previous implementations [HUA 07a, HUA 08, HUA 09], in the current work this stabilization scheme is *invoked* only when the required material usage volume V_{req} is reached, i.e. toward the end of the optimization loop.

In order to stabilize the design process in dealing with nonlinear dissipative structures, the basic BESO update scheme presented in section 1.4.3 is further enhanced with an introduction of two additional threshold parameters $\alpha_{\text{del}}^{\text{th}}$ and $\alpha_{\text{add}}^{\text{th}}$ for material removal and addition, respectively,

$$\rho_e^{(l+1)} = \begin{cases} 0 & \text{if } \alpha_e \leq \alpha_{\text{del}}^{\text{th}} \text{ and } \rho_e^{(l)} = 1, \\ 1 & \text{if } \alpha_{\text{add}}^{\text{th}} < \alpha_e \text{ and } \rho_e^{(l)} = 0, \\ \rho_e^{(l)} & \text{otherwise.} \end{cases} \quad [3.28]$$

The present scheme indicates that solid elements are removed when their sensitivity numbers are less than $\alpha_{\text{del}}^{\text{th}}$ and void elements are recovered when their sensitivity numbers are greater than $\alpha_{\text{add}}^{\text{th}}$. The parameters $\alpha_{\text{del}}^{\text{th}}$ and $\alpha_{\text{add}}^{\text{th}}$ are obtained from the following iterative algorithm:

1) let $\alpha_{\text{add}}^{\text{th}} = \alpha_{\text{del}}^{\text{th}} = \alpha_{\text{th}}$, where the value α_{th} is determined iteratively such that the required material volume usage is met at the current iteration;

2) compute the admission ratio c_{ar}, which is defined as the volume of the recovered elements divided by the total volume of the current design iteration. If $c_{\text{ar}} \leq c_{\text{ar}}^{\max}$, the maximum admission ratio, then skip the next steps; otherwise, $\alpha_{\text{del}}^{\text{th}}$ and $\alpha_{\text{add}}^{\text{th}}$ are redetermined in the next steps;

3) determine $\alpha_{\text{add}}^{\text{th}}$ iteratively using only the sensitivity numbers of the void elements until the maximum admission ratio is met, i.e. $c_{\text{ar}} \approx c_{\text{ar}}^{\max}$;

4) determine $\alpha_{\text{del}}^{\text{th}}$ iteratively using only the sensitivity numbers of the solid elements until the required material volume usage is met at the current iteration.

The introduction of $c_{\text{ar}}^{\max}$ stabilizes the topology optimization process by controlling the number of recovered elements. Normally, $c_{\text{ar}}^{\max}$ is set to a value greater than 1% so that it does not suppress the merit of the element recovery scheme.

3.2. Microscopic modeling

In this section, the considered material microstructure and the nonlinear constitutive law are discussed first in section 3.2.1. The computational complexity of the implementation of the considered microscopic model within the topology optimization framework is estimated in section 3.2.2. The basic concepts of pRBMOR-based model reduction for microscopic material modeling with GPU acceleration are briefly reviewed in 3.2.3. To be consistent with the source papers [FRI 15a, FRI 15b], we apply the exact notation rules for the presentation of the pRBMOR method.

3.2.1. *Elastoviscoplastic porous material*

A porous metal made of a ductile, elastoviscoplastic matrix with properties related to aluminum is considered. The periodic RVE used in [FRI 15a] containing 20 spherical pores with identical radii is considered (see also Figure 3.2). The porosity of the RVE is 10%. The FE discretization consists of a total of 140,508 nodes and 90,940 quadratic tetrahedral elements. An elastoviscoplastic overstress model based on a von Mises yield criterion is used for the ductile matrix. It fits into the generalized standard material framework (GSM, [HAL 75]), which is a key requirement of the model reduction technique proposed in section 3.2.3. Every GSM is characterized by two potentials: the Helmholtz free energy density ψ and the dissipation potential ϕ. Alternatively to ϕ, its Legendre transform ϕ^* referred to as the dual dissipation potential can be used.

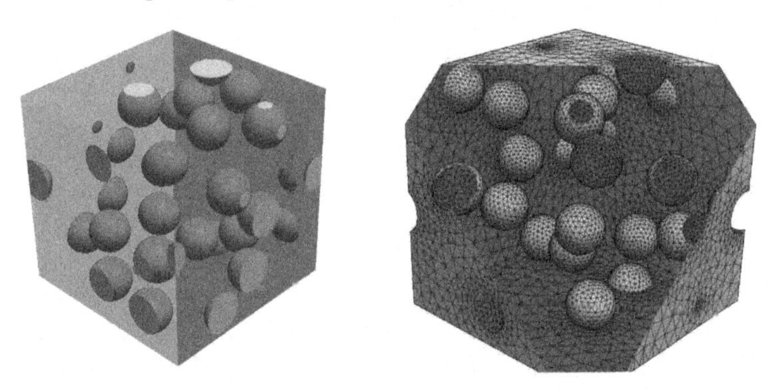

Figure 3.2. *Periodic RVE (left: geometry; right: finite element discretization). For a color version of the figure, see www.iste.co.uk/xia/topology.zip*

In the small strain framework considered in this work, the free energy depends on the total strain ε, the plastic strain tensor ε^{p} and the isotropic hardening variable q. The additive split

$$\psi(\varepsilon, \varepsilon^{\mathrm{p}}, q) = \psi_{\mathrm{e}}(\varepsilon^{\mathrm{e}}) + \psi_{\mathrm{h}}(q) \qquad [3.29]$$

into an elastic strain energy ψ_e depending on the elastic strain $\boldsymbol{\varepsilon}^\mathrm{e} = \boldsymbol{\varepsilon} - \boldsymbol{\varepsilon}^\mathrm{p}$ and the hardening potential ψ_h is considered. For the present application, Young's modulus $E = 75$ GPa and Poisson's ratio $\nu = 0.3$ are assumed. The strain energy then reads

$$\psi_\mathrm{e}(\boldsymbol{\varepsilon}^\mathrm{e}) = \frac{1}{2}\boldsymbol{\varepsilon}^\mathrm{e} \cdot \mathbb{C}[\boldsymbol{\varepsilon}^\mathrm{e}], \qquad \mathbb{C} = \frac{E}{3(1-2\nu)}\boldsymbol{I} \otimes \boldsymbol{I} + \frac{E}{1+\nu}\left(\mathbb{I}^\mathrm{s} - \frac{1}{3}\boldsymbol{I} \otimes \boldsymbol{I}\right). \tag{3.30}$$

The yield stress σ_F is defined as a function of the isotropic hardening variable q. It consists of an exponentially saturating part and a linear hardening contribution via (see also [FRI 15a])

$$\begin{aligned} \sigma_\mathrm{F}(q) &= \sigma_0 - r(q), \\ r(q) &= -\frac{\partial \psi_\mathrm{h}(q)}{\partial q} = -\left(h_\infty q + \sigma_\mathrm{h}^\infty \left[1 - \exp\left(\frac{h_\infty - h_0}{\sigma_\mathrm{h}^\infty} q\right)\right]\right). \end{aligned} \tag{3.31}$$

The static variable r is referred to as the thermodynamic conjugate to q. The parameters $h_\infty = 400$ MPa, $h_0 = 10,000$ MPa and $\sigma_0 = \sigma_\mathrm{h}^\infty = 100$ MPa were used. Note that the hardening behavior described by these parameters is highly nonlinear (Figure 3.3). The yield function is defined as

$$F(\boldsymbol{\sigma}, r) = \|\mathrm{dev}(\boldsymbol{\sigma})\| - \sqrt{\frac{2}{3}}\left(\sigma_0 - r(q)\right). \tag{3.32}$$

The material is elastic for $F < 0$ and elastoviscoplastic for $F > 0$. In the particular case of $F = 0$, a plastically neutral load state on the yield surface is defined, i.e. no inelastic processes take place.

In the following, we focus on a formulation of the dissipative process as a function of the thermodynamic driving forces by using the dual dissipation potential ϕ^*. The evolution of the internal variables (namely of the plastic strain tensor $\boldsymbol{\varepsilon}^\mathrm{p}$ and of the hardening variable q) is then determined by the dual dissipation potential

$$\phi^*(\boldsymbol{\sigma}, r) = \sqrt{\frac{2}{3}}\frac{\dot{\varepsilon}_0 \sigma_\mathrm{D}}{n+1}\left(\frac{\max\{F(\boldsymbol{\sigma}, r), 0\}}{\sqrt{2/3}\,\sigma_\mathrm{D}}\right)^{n+1} \tag{3.33}$$

via the relations

$$\dot{\varepsilon}^{\mathrm{p}} = \frac{\partial \phi^*(\boldsymbol{\sigma}, r)}{\partial \boldsymbol{\sigma}}, \qquad \dot{q} = \frac{\partial \phi^*(\boldsymbol{\sigma}, r)}{\partial r}. \qquad [3.34]$$

The parameters $n = 10$, $\dot{\varepsilon}_0 = 0.01$ s^{-1} and $\sigma_{\mathrm{D}} = 25$ MPa were chosen in analogy to [FRI 15a]. The viscoplastic behavior described by the proposed model mimics the behavior of ductile aluminum.

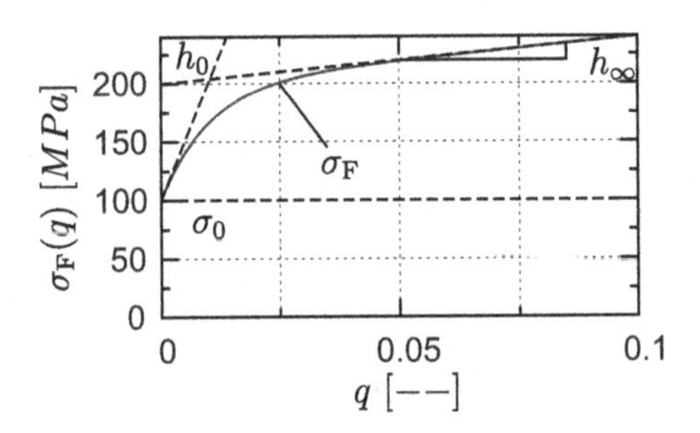

Figure 3.3. *Nonlinear yield stress as a function of the hardening variable* q

3.2.2. *Estimation of the computational complexity*

The major part of the computational cost of multiscale simulations involving nonlinearities is represented by the repetitive solution of nonlinear problems on the microscale. This concerns both CPU time and memory demands. In the particular context of multiscale topology optimization, the overall CPU time T_{CPU} and memory cost C_{mem} can be estimated as

$$T_{\mathrm{CPU}} = T_{\mu}\, \mathcal{O}(n_{\mathrm{top}} n_{\mathrm{gp}} n_{\mathrm{load}} n_{\mathrm{iter}}^{\mathrm{eq}}), \qquad C_{\mathrm{mem}} = n_{\mathrm{data}}^{\mu}\, \mathcal{O}(n_{\mathrm{gp}}), \qquad [3.35]$$

with the number of topology iterations n_{top}, the number of integration points of the macroscale problem n_{gp}, the number of load increments n_{load}, the average number of (macroscopic) nonlinear equilibrium iterations per load increment $n_{\mathrm{iter}}^{\mathrm{eq}}$ and the amount of data per microscale problem n_{data}^{μ}. An important factor is the average CPU time for the solution of one (equilibrated) load increment of the microscale problem T_{μ}, which includes the time for iterations on the RVE level in order to attain the equilibrium state in the presence of nonlinearities. Both T_{μ} and n_{data}^{μ} depend strongly on the spatial discretization of Ω_{μ}. Unfortunately, the discretization level of the microstructure can usually not be overly coarse due the geometric complexity of heterogeneous materials and the required accuracy of the microscale simulation. Similarly, the other influence factors (n_{top}, n_{gp} and $n_{\mathrm{iter}}^{\mathrm{eq}}$, n_{load}) are merely modifiable. In order to obtain a reasonable overall reduction of T_{CPU} and C_{mem}, it is, thus, indispensable to reduce both the microscopic solution time T_{μ} and the amount of microscopic information n_{data}^{μ}, significantly, i.e. by several orders of magnitude, without changing the spatial discretization. Model order reduction can be an appropriate tool for this.

3.2.3. *Model reduction using the pRBMOR*

In order to reduce computational effort by several orders of magnitude, the authors have recently proposed a ROM for GSMs: the pRBMOR [FRI 13b] that makes use of ideas originating from the NTFA [MIC 03, MIC 04, FRI 10]. The pRBMOR was extended in [FRI 14] in order to account for heterogeneously distributed hardening states on the microscopic scale. In the same work, algorithmic considerations have shown that the pRBMOR can be parallelized on many core systems such as GPUs. A specific GPU library was developed within Nvidia's compute unified device architecture (CUDA) framework. It was shown that the pRBMOR with the novel high-performance GPU implementation attains an overall reduction of the CPU time and, simultaneously, of the memory requirements on the order of 10^4 and beyond [FRI 14, FRI 15a].

Importantly, the mechanical predictions of the ROM show a sufficient accuracy with respect to the local and the effective stress predictions. Thereby, the pRBMOR qualifies as a tool for the application in nonlinear multiscale topology optimization, where classical computational homogenization methods [FEY 00, KOU 01, MIE 02] would lead to unacceptable numerical costs. The method has successfully been applied in FE^2 in terms of the FE^{2R} (FE Square Reduced) method [FRI 15a]. In the following, we briefly summarize the key ideas of the pRBMOR approach. For further algorithmic and theoretical details, we refer to [FRI 13b, FRI 14, FRI 15a].

The key idea in the pRBMOR is the utilization of a reduced basis for the internal variables, i.e. for the plastic strain and for the hardening variables. This is a major difference to other mechanical model reduction techniques that instead aim at a reduced representation of the displacement field. In the pRBMOR, the reduced basis is extracted from preceding nonlinear simulations using a snapshot POD and the respective basis functions are called modes. In the reduced setting, the approximations of the internal state are expressed via

$$\boldsymbol{\varepsilon}^{\mathrm{p}}(\boldsymbol{x},t) \sim \hat{\varepsilon}^{\mathrm{p}}(\boldsymbol{x},t) \approx \hat{P}(\boldsymbol{x})\hat{\xi}(t), \qquad \hat{q}(\boldsymbol{x},t) \approx \hat{Q}(\boldsymbol{x})\hat{\lambda}(t). \qquad [3.36]$$

The operators $\hat{P}(\boldsymbol{x})$ and $\hat{Q}(\boldsymbol{x})$ (referred to as relocalization operators) are columnwise composed of the plastic modes $\hat{\mu}^{(\alpha)}(\boldsymbol{x})$ and of the hardening modes $\hat{q}^{(\beta)}(\boldsymbol{x})$, respectively. Based on micromechanical ideas going back to the transformation field analysis [DVO 92, DVO 94], it can be shown that the strain and stress fields

$$\boldsymbol{\varepsilon}(\boldsymbol{x},t) \sim \hat{\varepsilon}(\boldsymbol{x},t) = \hat{E}_{\mathrm{e}}(\boldsymbol{x})\hat{\bar{\varepsilon}}(t) + \hat{E}_{\mathrm{p}}(\boldsymbol{x})\hat{\xi}(t) \qquad [3.37]$$

$$\boldsymbol{\sigma}(\boldsymbol{x},t) \sim \hat{\sigma}(\boldsymbol{x},t) = \hat{S}_{\mathrm{e}}(\boldsymbol{x})\hat{\bar{\varepsilon}}(t) + \hat{S}_{\mathrm{p}}(\boldsymbol{x})\hat{\xi}(t) \qquad [3.38]$$

solve the microscopic mechanical problem for arbitrary macroscopic strain $\bar{\boldsymbol{\varepsilon}}$ and mode activity $\hat{\xi}$. The operators $\hat{E}_{\mathrm{e}}$, $\hat{E}_{\mathrm{p}}$, $\hat{S}_{\mathrm{e}}$ and $\hat{S}_{\mathrm{p}}$ are obtained from linear precomputations in the presence of eigenstress fields induced by the plastic modes $\hat{\mu}^{(\alpha)}(\boldsymbol{x})$ and from linear elastic homogenization. Hence, all microscopic fields can be expressed as functions of the macroscopic variables $\hat{\bar{\varepsilon}}$, $\hat{\xi}$ and $\hat{\lambda}$. Making use of these local fields and introducing the effective thermodynamic driving forces $\hat{\tau}$ (for $\hat{\xi}$) and $\hat{R}$ (for $\hat{\lambda}$), an effective mixed incremental potential

$$\bar{\Pi}^*(\hat{\xi},\hat{\lambda},\hat{\tau},\hat{R};\hat{\bar{\varepsilon}}) = \langle\psi\rangle - \langle\psi_n\rangle + \Delta\hat{\xi}\cdot\hat{\tau} + \Delta\hat{\lambda}\cdot\hat{R} - \Delta t\,\langle\phi^*\rangle \qquad [3.39]$$

is defined. Here, ψ_n is the free energy evaluated at time t_n, $\Delta\hat{\xi}$ and $\Delta\hat{\lambda}$ are the increments of the macroscopic reduced variables and all other terms are evaluated at the current time $t = \Delta t + t_n$. Although $\bar{\Pi}^*$ depends heavily on $\hat{\bar{\varepsilon}}$, it shall be emphasized that $\hat{\bar{\varepsilon}}$ is an external control parameter. Following the derivations in [FRI 13b, FRI 14], the saddle point of $\bar{\Pi}^*$ implies a root of the nonlinear function

$$\hat{f}(\hat{\xi},\hat{\lambda},\hat{\tau},\hat{R}) \stackrel{!}{=} \hat{0} = \begin{bmatrix} \hat{\tau} - \hat{A}\hat{\bar{\varepsilon}} - \hat{D}\hat{\xi} \\ \hat{R} + \left\langle \hat{Q}^{\mathrm{T}}\partial_q\psi \right\rangle \\ \Delta\hat{\xi} - \Delta t\,\hat{D}^{-1}\left\langle \hat{S}_{\mathrm{p}}^{\mathrm{T}}\partial_\sigma\phi^* \right\rangle \\ \Delta\hat{\lambda} - \Delta t\,\hat{\bar{\mathcal{H}}}^{-1}\left\langle \hat{Q}^{\mathrm{T}}\hat{H}\partial_r\phi^* \right\rangle \end{bmatrix}, \qquad [3.40]$$

emerging from the necessary conditions of the inf-sup-condition. Here, the auxiliary matrices are

$$\hat{D} = -\left\langle \hat{S}_{\mathrm{p}}^{\mathrm{T}} \hat{C}^{-1} \hat{S}_{\mathrm{p}} \right\rangle,$$

$$\hat{A} = -\left\langle \hat{S}_{\mathrm{p}}^{\mathrm{T}} \right\rangle, \quad \hat{H} = \frac{\partial \hat{r}}{\partial \hat{q}} = -\frac{\partial^2 \psi}{\partial \hat{q}\, \partial \hat{q}}, \quad \hat{\bar{\mathcal{H}}} = \left\langle \hat{Q}^{\mathrm{T}} \hat{H} \hat{Q} \right\rangle, \qquad [3.41]$$

defining the mode interaction matrix $\hat{D}$, the strain sensitivity of the modes $\hat{A}$ and the local and the global hardening matrices $\hat{H}$ and $\hat{\bar{\mathcal{H}}}$, respectively. The solution of [3.40] yields the increments of the reduced variables $\Delta\hat{\xi}$, $\Delta\hat{\lambda}$ and the respective driving forces $\hat{\tau}$, $\hat{R}$. The effective stress can then be computed from the linear relation

$$\hat{\bar{\sigma}} = \hat{\bar{C}}\hat{\bar{\varepsilon}} - \hat{A}^{\mathrm{T}}\hat{\xi}, \qquad [3.42]$$

with $\hat{\bar{C}}$ being the homogenized linear elastic stiffness matrix.

3.2.4. *Implementation of pRBMOR*

With the abovementioned reduction strategy, the microscopic solution in the multiscale design framework (algorithm 1.1) can now be substituted by pRBMOR with GPU acceleration. Algorithm 3.1 presents the off-line phase for the initialization of pRBMOR. The on-line phase for the implementation of pRBMOR on GPUs is outlined in algorithm 3.2. The overall load is prescribed in several load steps. During each load step, several NR iterations are performed in order to meet the static equilibrium conditions on the structural/macroscopic scale, i.e. the design scale. At each of these iterations, we compute at each of the n_{gp} macroscopic Gauss integration points the strain $\bar{\varepsilon}$. Then, the current state consisting of the current strain $\bar{\varepsilon}$ and the internal variables $\hat{\boldsymbol{\xi}}$, $\hat{\boldsymbol{\lambda}}$ are written to a binary file (usually located on a RAM disc although I/O times have a small impact on the overall computing times). In the next step, the computing procedure is called for all n_{gp} integration points simultaneously using an external C/C++ program. The results of the computing procedure are the effective stress $\bar{\boldsymbol{\sigma}}$, the algorithmic tangent operator $\bar{\mathbb{C}}^{\mathrm{tan}}$ and the new reduced states $\hat{\boldsymbol{\xi}}$, $\hat{\boldsymbol{\lambda}}$. These variables are written to a binary file, which is then further processed at the macroscopic scale in order to assemble the global stiffness matrix and the nodal force residual. Note that in order to apply the pRBMOR in two-scale simulations, a robust time integrator is required as well as an accurate algorithmic tangent stiffness matrix. In the following, the pRBMOR is used based on the FE^{2R} implementation with GPU acceleration proposed in [FRI 15a]. A homotopy method is applied to improve the robustness of the time integration procedure and to obtain a meaningful algorithmic tangent operator.

Algorithm 3.1. Preanalysis for pRBMOR (off-line phase)

1: define the RVE model;
2: solve the RVE problem for different loadings;
3: compute correlation matrix and extract the reduced bases;
4: perform linear eigenstress analysis;
5: extract system matrices and pRBMOR data;

Algorithm 3.2. RVE solutions on GPUs (on-line phase)

1: read data of all macro Gauss points;
2: read pRBMOR data: $\bar{\varepsilon}, \hat{\boldsymbol{\xi}}, \hat{\boldsymbol{\lambda}}, \Delta\bar{\varepsilon}$;
3: initialize GPU: allocate memory and load data;
4: **loop** over all Gauss points
5: relocate microscopic state;
6: compute residual and Jacobian on the GPU;
7: iterate reduced variables $\hat{\boldsymbol{\xi}}, \hat{\boldsymbol{\lambda}}$;
8: **end loop**
9: **return** residual, stiffness matrices and $\hat{\boldsymbol{\xi}}, \hat{\boldsymbol{\lambda}}$.

3.3. Numerical examples

A cantilever design problem similar to the one studied in [BEN 03] is considered. The two-dimensional problem setting is illustrated in Figure 3.4. The elastoviscoplastic RVE presented in section 3.2.1 is assumed at the microscopic scale. The dimension of the cantilever is assumed to be 2 m × 1 m. The left end of the cantilever is clamped and displacement loading at a portion of the right-end edge (25%) is prescribed.

Two variations of the cantilever problem are investigated: first, the BESO method is used in the absence of dissipative effects but with consideration of the effective elastic properties of the microstructure (see section 3.3.1). The linear designs are computed in order to first investigate the properties of the BESO for the elastic design process and in order to gather information regarding the coarsest possible mesh resolution that is sufficient to provide meaningful results at balanced computational expense. Then, nonlinear topology optimization using the new BESO-FE^{2R} approach are discussed in section 3.3.2. The optimal topologies obtained in the linear optimization are used in nonlinear simulations in order to quantify the improvements of the structural design due to application of the nonlinear BESO-FE^{2R} approach (see section 3.3.4).

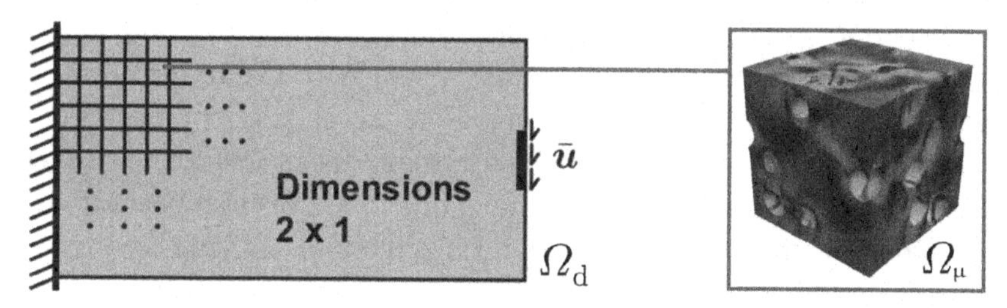

Figure 3.4. *Illustration of a two-scale cantilever structure with associated porous microstructure. For a color version of the figure, see www.iste.co.uk/xia/topology.zip*

Square shaped bilinear elements are used for the FE discretization of the cantilever and plane strain conditions are assumed. For the elastic case (BESO), various mesh resolutions ranging from 32×16 up to 256×128 elements are considered in order to examine convergence of the objective function with respect to mesh refinements. For the inelastic case (BESO-FE^{2R}), two different mesh densities are considered (see section 3.3.2 for details). All outputs (e.g. forces, energies) are specified relative to the depth d of the cantilever, i.e. per unit length.

In the following, all parameters involved in the BESO method are held constant for all topology optimizations except for $c_{\rm er}$. The evolution rate, which determines the percentage of removed material at each design iteration, is set to $c_{\rm er} = 0.01$ for the linear optimization. For the nonlinear design problem, the values 0.01 and 0.02 are considered. The maximum admission ratio corresponding to the maximum percentage of recovered material that is allowed per iteration is set to $c_{\rm ar}^{\rm max} = c_{\rm er}$. In the following, $n_{\rm y}$ denotes the number of elements in vertical direction and $n_{\mathrm{x}} = 2n_{\mathrm{y}}$ is the mesh resolution in horizontal direction. The filter radius is set to $r_{\min} = n_{\mathrm{y}}/16\, l^{\mathrm{e}}$ in each of the following linear and nonlinear simulations.

3.3.1. *Linear elastic topology optimization*

For the linear elastic topology optimization, the homogenized stiffness of the porous microstructure is used at each macroscopic integration point of the design domain that is associated with the solid (microheterogeneous) material. By using the effective stiffness $\bar{\mathbb{C}}$, the microstructural geometry and the distribution of the material parameters within the microstructure are captured at a negligible computational investment. More precisely, the evaluation of $\bar{\mathbb{C}}$ requires the solution of the linear RVE problem for six linearly independent load cases once in the preprocessing. While the effective stiffness tensor is not completely isotropic due to the random position of the pores, approximations of the effective bulk modulus $\bar{K}$ and of the shear modulus $\bar{G}$ can be computed by means of projecting $\bar{\mathbb{C}}$ onto the two isotropic projectors

$$3\bar{K} = \mathbb{P}_1 \cdot \bar{\mathbb{C}} = \frac{1}{3}\boldsymbol{I} \otimes \boldsymbol{I} \cdot \bar{\mathbb{C}}, \qquad 2\bar{G} = \frac{1}{5}\mathbb{P}_2 \cdot \bar{\mathbb{C}} = \frac{1}{5}\left(\mathbb{I}^{\mathrm{s}} - \mathbb{P}_1\right) \cdot \bar{\mathbb{C}}. \tag{3.43}$$

The deviation of $\bar{\mathbb{C}}$ from its isotropic approximation is on the order of 0.5%, i.e. in the linear regime almost isotropic properties are predicted. Therefore, near symmetric topologies are expected. A tip deflection of 10 mm is applied.

The linear design process for the considered two-dimensional problem is computationally rather inexpensive and can be performed on a standard laptop computer using a Matlab-based implementation. Figure 3.5 outlines the converged solutions for the linear elastic porous material (i.e. in the absence of dissipative effects) for three mesh resolutions ranging from 64×32 to 256×128 elements. The volume fraction of the solid (microporous) material is 60%.

Convergence was attained after 42–54 topology iterations depending on the mesh resolution, respectively. Although these numbers depend on the convergence criterion, they give an indication on the number of required optimization cycles. Notably, the number of iterations was almost independent of the discretization (i.e. of the number of design variables). For all mesh densities, dissymmetric final designs are observed. The degree of the dissymmetry is surprising when considering the marginal anisotropy of the effective linear elastic material behavior. A closer investigation of the design iterations showed that the slight anisotropy of the material results in minor dissymmetries during the initial design iterations. These are then conserved and/or amplified during the optimization.

Figure 3.5. *Design solutions for linear elasticity for the three mesh resolutions* 64×32, 128×64 *and* 256×128 *elements (left to right; macroscopic volume fraction 60%)*

The objective function f_w determined during the automatic topology optimization is plotted over the number of elements along the vertical direction n_y in Figure 3.6, stating only minor variations with respect to varying mesh resolution. The physical dimension of f_w is kJ/mm due to the dependence of the overall strain energy on the depth d of the structure. The small variation of the designs with respect to variations of the FE discretization confirms that the chosen BESO with stabilization is robust and that the related parameters are well calibrated. It is found that $n_\mathrm{y} = 16$ elements in the vertical direction might not be sufficient due to overly pronounced discretization errors in the macroscopic FE problem. Therefore, $n_\mathrm{y} = 32$ and 64 are investigated in the nonlinear topology optimization, i.e. 64×32 and 128×64 elements are considered.

3.3.2. *Nonlinear structural design using FE^{2R}*

In the following, the same tip deflection of 10 mm is applied, but instead of using the homogenized linear properties of the porous material, the FE$^{2\mathrm{R}}$ method is used in order to account for the viscoplasticity of the solid material with consideration of the highly nonlinear hardening law. Based on the results of the linear elastic topology optimization, a mesh resolution of $n_\mathrm{y} = 64$ elements in vertical direction (i.e. $128 \times 64 = 8192$ elements in total) is considered as the maximum mesh density for the nonlinear two-scale optimization. Additionally, $n_y = 32$ is considered for computationally less expensive comparison designs in order to investigate the influence of different parameters of the optimization procedure as well as the influence of the mesh density in the presence of nonlinearities.

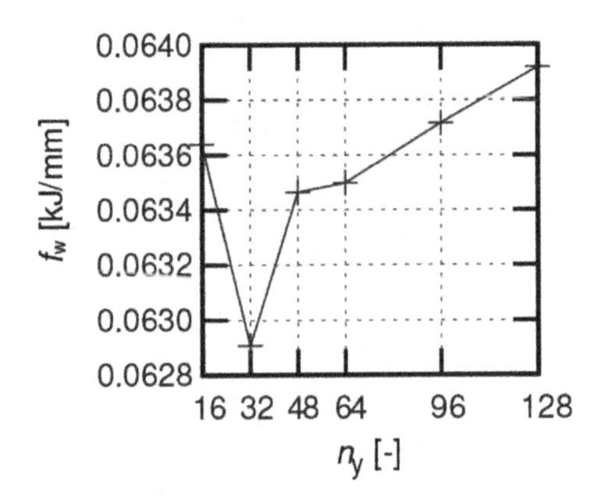

Figure 3.6. *Objective function f_w for the linear design problem as a function of the number of elements in y-direction n_y (with $n_x = 2n_y$)*

The impact of a variation of the parameter c_{er} is studied for both mesh resolutions. Therefore, the BESO-FE^{2R} algorithm was applied using $c_{er} = 0.01$ and 0.02. The intention of this comparison is to investigate the effect of the step size used in the topology optimization. During the topology iterations, the volume fraction of the solid is continuously reduced, i.e. after each nonlinear FE simulation on the macroscale, the current target volume fraction is reduced until the prescribed volume fraction is attained. Only then, pure rearrangements of the existing solid are performed in order to reduce the objective function. Another option producing a distinctly increased computational cost would be to converge for each of the intermediate volume fractions $c^{(l)}$ to the respective optimal design before further reduction of c. Since the computational expense for also converging the intermediate designs is very large, small values of c_{er} are investigated. The thereby induced smaller adjustments of the intermediate solid volume fractions lead to almost converged intermediate designs. They could help to understand the effect of converged intermediate designs on the final result of the optimization process. Four optimizations yield the following objective values:

– $f_{\mathrm{w}}(c_{\mathrm{er}} = 0.02, n_{\mathrm{y}} = 32) = 0.05839$ kJ/mm after 26 iterations;

– $f_{\mathrm{w}}(c_{\mathrm{er}} = 0.01, n_{\mathrm{y}} = 32) = 0.05854$ kJ/mm after 41 iterations;

– $f_{\mathrm{w}}(c_{\mathrm{er}} = 0.02, n_{\mathrm{y}} = 64) = 0.05861$ kJ/mm after 27 iterations;

– $f_{\mathrm{w}}(c_{\mathrm{er}} = 0.01, n_{\mathrm{y}} = 64) = 0.05843$ kJ/mm after 45 iterations.

These values suggest that: (i) the mesh resolution, and (ii) the volume fraction change per design iteration c_{er} are both of minor importance for the inelastic design.

For both mesh densities, it can be observed in Figure 3.7 that drops of the objective function under an increment of the void volume fraction ρ_{V} occur. This is due to sudden and, at the same time, distinct topological changes such as disappearing rod-like connectors. An example for such a sudden change in the topology is illustrated in Figure 3.8, for $n_{\mathrm{y}} = 64$, $c_{\mathrm{er}} = 0.02$ with 17–21 iterations. The void volume fraction grows linearly from 32% (it. 17) to 40% (it. 21 onwards). The values of the objective function reported in Figure 3.7 confirm that – despite further material removal – the stiffness of the structure increases after iteration 18. This increase is due to a favorable rearrangement of the remaining material.

The four final designs are compared in Figure 3.9. It is found that all designs tend to be dissymmetric. Interestingly, the topologies obtained for $c_{\mathrm{er}} = 0.01$ are almost mesh independent. A detailed comparison of the two designs for $c_{\mathrm{er}} = 0.01$ is carried out in Figure 3.10. It is found that the designs have 89.9% of the solid material in common, which confirms that the two designs differ only moderately.

In order to solve the four two-scale design problems provided in this section alone, a total of 139 two-scale FE problems were simulated involving the solution of approximately 30 mio. load increments on the RVE level. Remarkably, these computations were conducted within little more than 1 week using a single GPU. These numbers clearly show the computational capability of the proposed technique and the impossibility to solve the same problem without sophisticated model reduction methods.

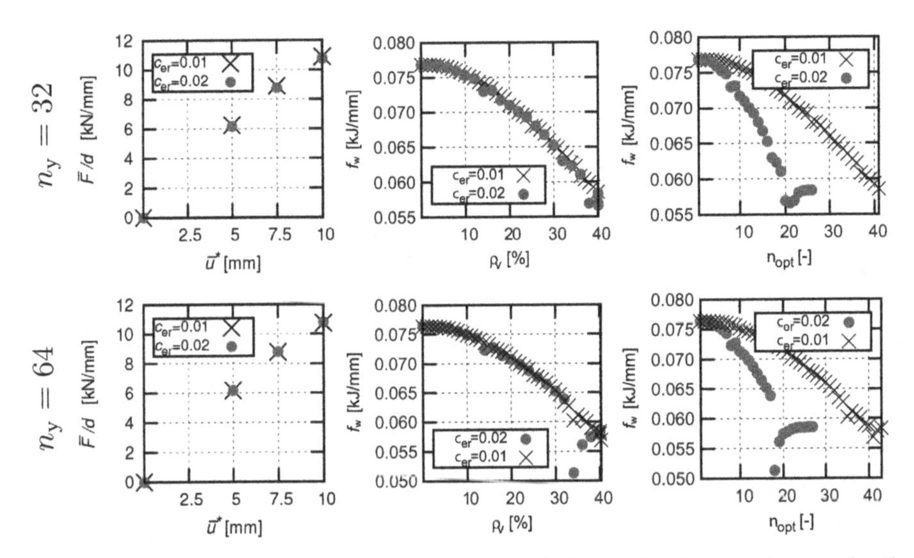

Figure 3.7. *Left: Comparison of the macroscopic force–displacement curve for the nonlinear designs for* $c_{\rm er} = 0.02$ *and* $c_{\rm er} = 0.01$*; middle, right: objective function versus the void volume fraction* $\rho_{\rm v}$ *(middle) and the number of topology iterations* $n_{\rm top}$ *(right); top:* $n_{\rm y} = 32$*; bottom:* $n_{\rm y} = 64$

Figure 3.8. *Topology iterations 17–21 (left to right) for* $n_{\rm y} = 64$, $c_{\rm er} = 0.02$*: an important topological change from 17* $\rightarrow$ *18 (two connecting rods disappear) requires several subsequent steps in order to recover good values of* Φ *(see bottom right graph in Figure 3.7)*

3.3.3. *Investigation of the influence of the load amplitude*

In addition to the already demanding multiscale topology optimizations presented before, additional computations were carried out in order to investigate the influence of the load amplitude on the final design. Therefore, all four optimizations (for the two different mesh densities and for the two different values of $c_{\rm er}$) were relaunched using seven instead of three load increments that lead to a final tip deflection of 20 mm instead of 10 mm. Figure 3.11 shows the final designs obtained. The computational effort for these optimizations is considerably increased over the previous computations.

Interestingly, the number of topology iterations n_{top} was found to be almost independent of the mesh size (26 ($n_{\text{y}} = 32$) versus 25 ($n_{\text{y}} = 64$) iterations $c_{\text{er}} = 0.02$, and 46 ($n_{\text{y}} = 32$) versus 44 ($n_{\text{y}} = 64$) iterations for $c_{\text{er}} = 0.01$. Moreover, the number of design iterations is almost the same as before despite the doubling of the tip deflection. The optimal objective functions obtained using the BESO-FE$^{2\text{R}}$ are

- $f_{\text{w}}(c_{\text{er}} = 0.02, n_{\text{y}} = 32) = 0.1872$ kJ/mm after 26 iterations;
- $f_{\text{w}}(c_{\text{er}} = 0.01, n_{\text{y}} = 32) = 0.1859$ kJ/mm after 46 iterations;
- $f_{\text{w}}(c_{\text{er}} = 0.02, n_{\text{y}} = 64) = 0.1859$ kJ/mm after 25 iterations;
- $f_{\text{w}}(c_{\text{er}} = 0.01, n_{\text{y}} = 64) = 0.1888$ kJ/mm after 44 iterations.

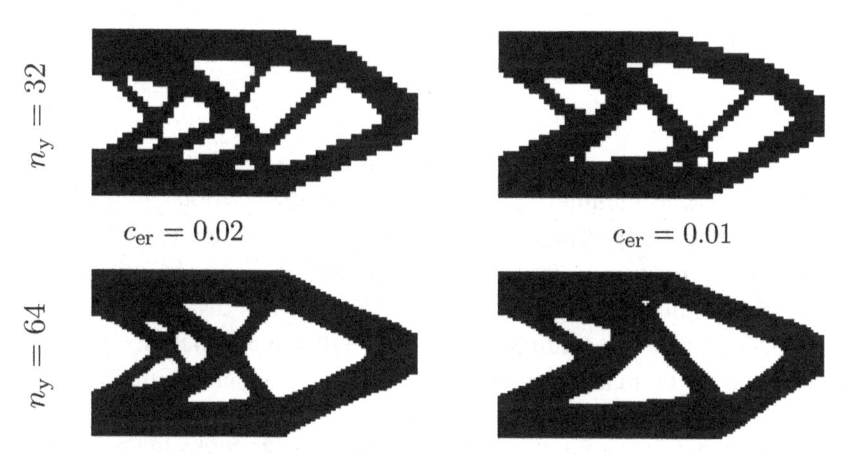

Figure 3.9. *BESO-FE^{2R} designs; $c_{\text{er}} = 0.02$ (left) and 0.01 (right); tip deflection $\bar{u} = 10$ mm*

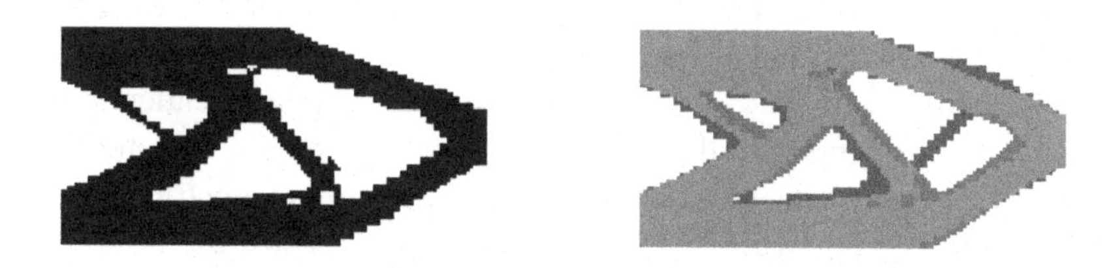

Figure 3.10. *Detailed design comparison for $c_{\text{er}} = 0.01$ for the different mesh densities; left: common region; right: additional material for $n_{\text{y}} = 32$ (magenta) and $n_{\text{y}} = 64$ (blue); tip deflection $\bar{u} = 10$ mm. For a color version of the figure, see www.iste.co.uk/xia/topology.zip*

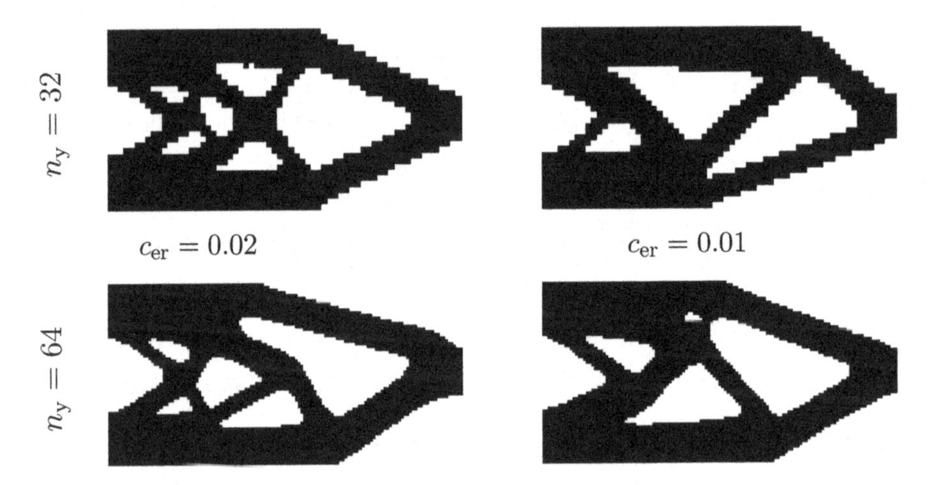

Figure 3.11. *BESO-FE^{2R} designs for the increased tip deflection of 20 mm for different parameters c_{er}, n_y*

The force–displacement curve and the objective function versus the (macroscopic) void volume fraction and number of iterations are reported in Figure 3.12. For the increased load amplitude, the nonlinearity of the overall mechanical response is much more pronounced. As in the previous investigation, sudden changes in the topology (such as disappearing rod-like connectors) induce discontinuities in the objective function, which are compensated for by material rearrangement in subsequent iterations.

For the increased loading, the four designs differ more than before. By closer examination, the two designs for the finer mesh are rather closely related having 88.8% of the solid elements in common. The design for $n_y = 32$, $c_{er} = 0.01$ (top right in Figure 3.11) is similarly related to the corresponding design of the finer mesh after mirroring with respect to the horizontal axis (with 88.4% of the solid elements being identical). From mechanical considerations, the mirroring of the mesh is a transformation leading to an equivalent mechanical design given the symmetric boundary conditions in the chosen linear kinematic framework. Thus, the designs can be considered to be in good agreement with each other that suggests a certain robustness of the BESO-FE^{2R} algorithm, although the resemblance could be doubted at first sight.

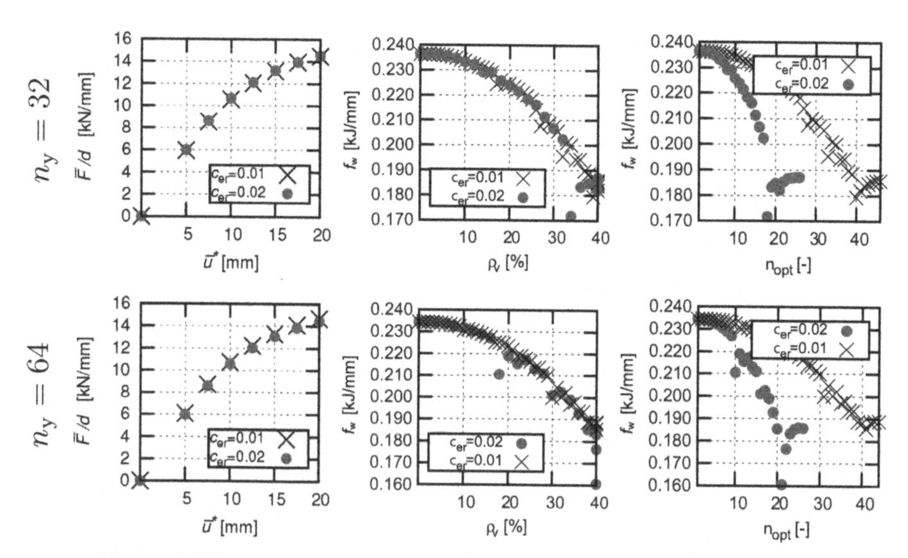

Figure 3.12. *Results of the optimization for the tip deflection of 20 mm; left: comparison of the macroscopic force–displacement curve for the nonlinear designs for $c_{\text{er}} = 0.02$ and $c_{\text{er}} = 0.01$; middle, right: objective function versus the void volume fraction ρ_{v} (middle) and the number of topology iterations n_{top} (right)*

Finally, the design for the fine mesh and $c_{\text{er}} = 0.01$ for the two different load amplitudes (10 and 20 mm tip deflection) are compared in Figure 3.13. Remarkably, the designs have 91.7% elements in common. Hence, the final design can be considered to be sufficiently robust with respect to moderate variations in the amplitude of the boundary conditions.

Figure 3.13. *Comparison of the two designs for the finer mesh; left to right: final design for $\bar{u} = 10$ mm, $\bar{u} = 20$ mm, comparison (gray: common, blue: additional for 10 mm, magenta: additional for 20 mm). For a color version of the figure, see www.iste.co.uk/xia/topology.zip*

3.3.4. *Comparison to linear designs*

Two different designs are investigated for the mesh consisting of 128×64 elements: (i) the design optimized using linear elasticity as discussed in section 3.3.1, and (ii) the design obtained using the nonlinear BESO-FE^{2R} (using the result of $c_{er} = 0.01$). The different geometries are subjected to the same loading (tip displacement of 20 mm at 25% of the right-hand side of the cantilever) and with consideration of the nonlinearity by using the FE^{2R}. The nonlinear two-scale problem is solved for the elastic design and the objective function is computed. For the elastic design, the objective function is 0.1839 kJ/mm while the BESO-FE^{2R} nonlinear design attains 0.1888 kJ/mm. Figure 3.14 shows the different topologies and the force–displacement curves of the two designs are compared. Geometrically, the linear design differs considerably from the nonlinear one. Since the mechanical response of the structure is mostly linear for the first half of the loading, the linear design can outperform the nonlinear design up to a tip deflection of 7.5 mm, which can be seen from the force displacement curve. However, the structure optimized for a tip deflection of 20 mm and consideration of the nonlinearity of the material behavior outperforms the topology optimized for elastic loading for higher tip deflections. Although the objective values are almost identical, the nonlinear design has major advantages in the second half of the loading, which is governed by the nonlinearity of the material behavior. Thereby, the chosen topology optimization criterion is validated. Note that the number of design iterations for the elastic and the inelastic design is almost identical for all considered loadings (47 for elastic versus 45 for 10 mm and 44 for 20 mm inelastic tip deflection, respectively).

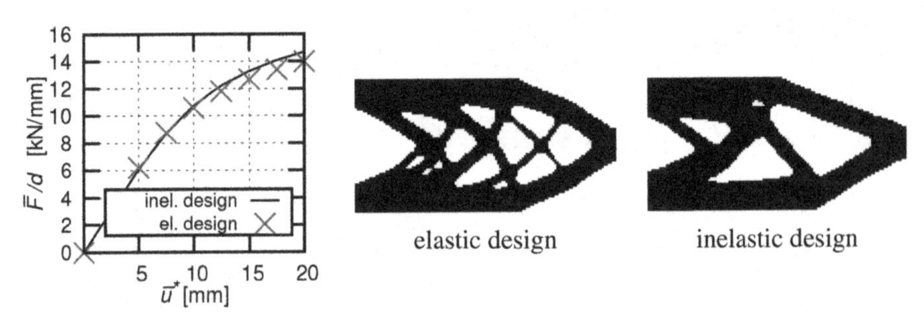

Figure 3.14. *Comparison of different designs: elastic BESO (left) versus nonlinear BESO-FE^{2R} design (right); macroscopic force–displacement curve (middle)*

3.4. Concluding remarks

In this chapter, we have extended the multiscale design framework for multiscale elastoviscoplastic structures with the implementations of the pRBMOR method and GPU parallelization. While the computational effort is still considerable, the use of pRBMOR renders the solution of the problem feasible on standard workstations with Nvidia GPUs. Using the FE^2 computing scheme directly without reduction, the design problem could only be solved at a computational investment that would be beyond current (and likely future) capabilities. Hence, the proposed method allows us to tackle problems that were until recently unanswerable at moderate computational expense.

The computational findings state that the optimal designs generally depend on the parameters chosen for the BESO method. However, the main topological and morphological features show certain invariance with respect to the parameters. This is quantified by the rather large amount of common elements (see section 3.3.2). The impact of the nonlinearity on the final design is most obvious when comparing the optimal elastic design to the inelastic design in terms of both their topological appearances and physical performances.

Although the designs are generally rather similar, the differences in the details of the designs are an indicator that further research for dissipative nonlinear topology optimization that could help to obtain more robust and efficient design schemes. During the optimization, sudden design changes having a major impact on the objective function are encountered. Possible extensions of the algorithm could include multiple design iterations at constant volume fraction of the solid phase in these cases. This could help to also improve the robustness of the design process for larger values of the evolutionary ratio c_{er}, thereby helping to further reduce the computation time.

The ultimate goal would be to simultaneously optimize the microstructure and the macroscopic design. The extension of the current multiscale design framework to simultaneous design of both the structure and the underlying microstructures could be realized by the induction of additional design variables at the microscopic scale. Nevertheless, the problem becomes extremely computationally time consuming due to the enormous number of suboptimization problems, especially for nonlinear cases. In the following chapter, we will start to extend the multiscale design framework for simultaneous topology optimization at both scales in linear elasticity.

4

Simultaneous Topology Optimization of Structure and Materials

We have so far assumed a fixed material microstructure at the microscopic scale. The ultimate goal of multiscale design would be to simultaneously optimize both the structure and the underlying microstructures. Designing materials simultaneously along with the design of structures would result in higher performance structures. In addition, the recently emerging and rapidly developing techniques of 3D printing provide the capability of manufacturing extremely fine and complex microstructures, which make it possible to generate more innovative, lightweight and structurally efficient designs.

In this chapter, we extend the multiscale design framework by introducing additional topology variables at the microscopic scale to perform simultaneous topology optimization of structure and material microstructures in the linear case. From the previously established multiscale design framework, we have topology variables and volume constraints defined at both scales. Cellular material models are defined in a pointwise manner. In this model, the material microstructures are optimized in response to the macroscopic solution, which results in the nonlinear equilibrium problem at the interface of the two scales. We treat the material optimization process integrally as a generalized nonlinear constitutive behavior, and the nonlinear scale-interface equilibrium problem can be resolved naturally within the multiscale framework by the FE^2 method.

As stated in [BEN 03] and tested by us [XIA 15a] (see the Appendix), the SIMP model is sensitive to the choice of initial guess, loadings and other parameters. The Appendix presents our recent work with educational purposes regarding design of materials using topology optimization and energy-based homogenization approach in Matlab. In contrast, the BESO method has shown a very robust and efficient performance in material designs for its heuristic design scheme [HUA 11, HUA 13, YAN 14]. In this work, we employ the BESO method for the designs at both scales.

This chapter is organized in the following manner. Section 4.1 reviews the general mathematical formulation of the simultaneous material and structural design. Section 4.2 presents the initial stiffness NR solution scheme. Section 4.3 gives the topology optimization models for both macro- and microscopic scale problems. Section 4.4 showcases the developed model by several numerical test examples. Concluding remarks are given in section 4.5

4.1. Problem statement and decomposition

Generalized mathematical formulations for simultaneous cellular material and structure designs can be found in [THE 99] and its application for continuous models has been given by [ROD 02]. Let x and y denote positions of a point at macroscopic and microscopic scales, respectively. The structural compliance minimization problem is stated in terms of two levels of design variables: the pointwise topology variable $\rho(x)$ at the macroscopic scale (structure) and the pointwise topology variable $\eta(x, y)$ at the microscopic scale (material).

Recalling [BEN 03], using the principle of minimum potential energy, the minimum compliance problem in a displacement-based formulation is given as:

$$\max_{(\rho,\eta)\in\mathcal{A}_{\mathrm{ad}}} \min_{u\in\mathcal{U}} \left\{ \frac{1}{2}\int_{\Omega} C_{ijkh}\left(x, \rho(x), \eta(x,y)\right) \frac{\partial u_i}{\partial x_j}\frac{\partial u_k}{\partial x_h} \mathrm{d}\Omega - l(u) \right\}. \qquad [4.1]$$

Here, $C_{ijkh}\left(x, \rho, \eta\right)$ is the fourth-order elastic stiffness tensor at material point x depending on both values of $\rho(x)$ and $\eta(x, y)$ at the two scales. $\mathcal{U}$ denotes the space of kinematically admissible displacement fields and $l(u)$ is the loading potential term. Note that although [4.1] is defined under a linear assumption, C_{ijkh} may depend in a nonlinear way on the design variables.

$\mathcal{A}_{\text{ad}}$ is the assembled admissible set of design variables and consists of two defined admissible sets $\mathcal{A}_\rho$ and $\mathcal{A}_\eta$ for $\rho(x)$ and $\eta(x, y)$, respectively,

$$\mathcal{A}_{\text{ad}} = \{\rho, \eta \mid \rho(x) \in \mathcal{A}_\rho, \eta(x, y) \in \mathcal{A}_\eta\}. \quad [4.2]$$

In the case of discrete topology design models [YAN 14, XIA 14a], $\mathcal{A}_\rho$ and $\mathcal{A}_\eta$ are simply defined as:

$$\mathcal{A}_\rho = \left\{\rho \mid \rho = 0 \text{ or } 1, \int_\Omega \rho(x) \mathrm{d}\Omega = V_{\text{req}}^{\text{s}}\right\}, \quad [4.3]$$

and

$$\mathcal{A}_\eta = \left\{\eta \mid \eta = 0 \text{ or } 1, \int_{\Omega_x} \eta(x, y) \mathrm{d}\Omega^x = V_{\text{req}}^x\right\}, \quad [4.4]$$

where $V_{\text{req}}^{\text{s}}$ and V_{req}^x are the allowed material volume at the macro- and microscales, respectively. Note that V_{req}^x can vary from point to point.

In the case of continuous topology optimization models (e.g., SIMP, [ROD 02, ZHA 06, NAK 13]), the elastic stiffness tensor and V_{req}^x for macroscale point x are functions of $\rho(x)$. In the current context, the discrete-valued $\rho(x)$ indicates only the existence of an additional fine scale ($\rho = 1$) or not ($\rho = 0$). We can therefore extract $\rho(x)$ outside C_{ijkh} and the remaining elastic stiffness tensor C_{ijkh} is dependent on $\eta(x, y)$, i.e.

$$\max_{(\rho,\eta) \in \mathcal{A}_{\text{ad}}} \min_{u \in \mathcal{U}} \left\{\frac{1}{2} \int_\Omega \rho(x)\, C_{ijkh}(x, \eta(x, y)) \frac{\partial u_i}{\partial x_j} \frac{\partial u_k}{\partial x_h} \mathrm{d}\Omega - l(u)\right\}. \quad [4.5]$$

The separation of the two-scale variables and the interchange of the equilibrium and local optimizations of [4.5] result in a reformulated displacement-based problem

$$\max_{\rho \in \mathcal{A}_\rho} \min_{u \in \mathcal{U}} \left\{\int_\Omega \bar{w}(x, u, \eta(x, y))\, \mathrm{d}\Omega - l(u)\right\}, \quad [4.6]$$

where the pointwise maximization of the strain energy density $\bar{w}$

$$\bar{w} = \max_{\eta \in \mathcal{A}_\eta} \frac{1}{2} \rho(x) C_{ijkh}(x, \eta(x, y)) \frac{\partial u_i}{\partial x_j} \frac{\partial u_k}{\partial x_h} \quad [4.7]$$

is treated as a subproblem for the prescribed $\rho(x)$ and $u(x)$ at the macroscale point x. From the reformulated form of [4.6], a hierarchical iterative solution strategy is straightforwardly established for the simultaneous material and structure design.

The outer maximization problem of [4.6] is the "master" problem dealing with the macroscale material distribution in terms of $\rho(x)$ for the macroscale structure. The inner maximization problems of [4.6], i.e. [4.7], are the "slave" problems corresponding to the stiffness maximizations of the microscale materials in terms of $\eta(x, y)$ for the evaluated macroscale strain. The middle layer minimization problem of [4.6] seeks kinematically admissible equilibrium displacements for the locally optimum energy function for the given distribution of the macroscale topology of $\rho(x)$. Note that since the locally optimum energies depend on the displacement field in a complex fashion via the optimization problems of [4.7], the equilibrium statement of [4.6] is in fact a constitutively nonlinear elastic problem.

4.2. Initial stiffness NR solution scheme

In order to solve the nonlinear scale-interface equilibrium problem of [4.6], the incremental computational homogenization approach (FE^2, [FEY 00], section 1.1) is used to bridge the two separated scales. Generally speaking, FE^2 solves two nested boundary value problems, one at the macroscopic scale and another at the microscopic scale. Because the constitutive behaviors can be implicitly represented by the stress–strain relationships, the determination of the homogenized elastic stiffness tensor of the optimized cellular microstructure is no longer required in the presented algorithm.

The material model defined at the microscopic scale is optimized upon the macroscale strain value at associated integration point. Then, the effective stress is evaluated on the optimized microscale material microstructure and returned to the upper scale. With the effective stress–strain relationship, scale-interface nonlinear equilibrium is then resolved by means of the NR method. The macroscale topology is then optimized using the converged solution. A schematic illustration is shown in Figure 4.1. Unlike in Chapters 1 and 2 where nonlinearities come from material nonlinear constitutive behaviors, the nonlinearity here is due to the optimization of microscopically defined materials. This nonlinearity can be viewed as a generalized nonlinear constitutive behavior. In summary, the FE^2-based solution scheme consists of the following steps:

1) evaluate the macroscale strain $\bar{\varepsilon}(\boldsymbol{x})$ with an initially defined setting;

2) define PBC on the associated material model according to $\bar{\varepsilon}(\boldsymbol{x})$;

3) optimize the topology η of the associated material model as indicated in [4.7];

4) evaluate the microscale stress $\boldsymbol{\sigma}(\boldsymbol{x}, \boldsymbol{y})$ on the optimized material topology η;

5) compute the macroscale stress $\bar{\boldsymbol{\sigma}}(\boldsymbol{x})$ via volume averaging $\boldsymbol{\sigma}(\boldsymbol{x}, \boldsymbol{y})$;

6) evaluate the consistent stiffness tensor $\bar{\mathbb{C}}(\boldsymbol{x})$ at macroscale point $\boldsymbol{x}$;

7) update the displacement solution using the NR method;

8) repeat steps 2–7 until the macroscale equilibrium is achieved.

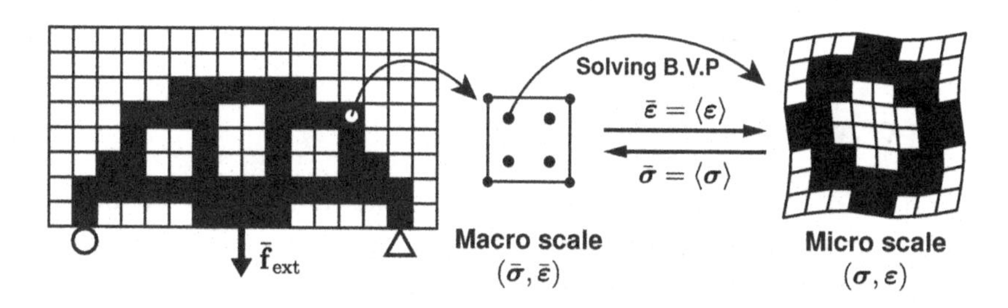

Figure 4.1. *FE^2-based simultaneous topology optimization of structure and materials [XIA 14a]*

Due to the particularity of the concerned nonlinearity, the conventional NR solution scheme using a tangent stiffness matrix is not applicable here. As can be observed in Figure 4.2, the tangent stiffness matrix for $\bar{\mathbf{u}}^{(1)}$ is in fact the linear stiffness matrix $\bar{\mathbf{K}}_{\text{opt}}(\bar{\mathbf{u}}^{(1)})$ itself. Using this stiffness matrix results in the divergence of the solution of the scale-interface nonlinear equilibrium. The solution of this type of nonlinearity itself is still an open issue according to the author's knowledge. We propose to use an initial stiffness NR solution scheme based on a reasonable hypothesis that the structure constituted by the optimized materials $(\bar{\mathbf{K}}_{\text{opt}}(\bar{\mathbf{u}}_{\text{sol}}))$ is stiffer than the other structures $(\bar{\mathbf{K}}_{\text{opt}}(\bar{\mathbf{u}}^{(1)}), \ldots)$ corresponding to the other admissible solutions. In this scheme, the applied initial stiffness matrix $\bar{\mathbf{K}}_0$ is constructed assuming the microscale material is full of solid material. Though not rigorous enough, this solution scheme is capable of dealing with this scale-interface nonlinearity with satisfactory as can be seen from numerical tests in section 4.4.

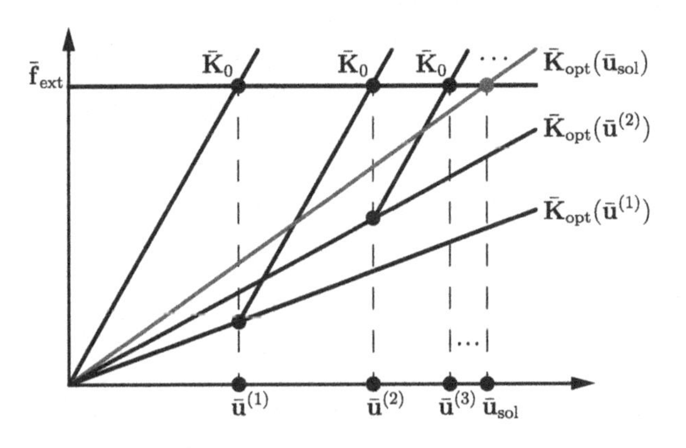

Figure 4.2. *Initial stiffness NR solution scheme [XIA 14a]*

4.3. Topology optimization models

The definitions of topology optimization model is given in section 4.3.1. The decomposed optimization models and corresponding sensitivity analysis are given in section 4.3.2. The BESO method is then briefly reviewed in section 4.3.3. The simultaneous design framework is then presented in section 4.3.4.

4.3.1. *Model definitions*

Within the context of FEA, both topology design variables

$$\boldsymbol{\rho} = (\rho_1, \ldots, \rho_{N_s})^T \quad i = 1, \ldots, N_s$$

$$\boldsymbol{\eta}^x = (\eta_1^x, \ldots, \eta_{N_x}^x)^T \quad j = 1, \ldots, N_x \qquad [4.8]$$

written in vector form, are defined in an elementwise manner at both scales. N_s and N_x are the numbers of discrete elements at the macro- and microscales, respectively. Here, the superscript x of $\boldsymbol{\eta}^x$ denotes a vector of microscale topology variables at each macroscale point x. Through discrete topology optimization, both types of variables take values of either 0 or 1. Note that in practice in order to prevent the singularity of the stiffness matrix, a small value is attributed to ρ_i or η_j to denote void elements. Consistently with the FE^2 computing scheme, each Gauss integration point x of the macroscopic structural mesh is attributed with a cellular material model $\boldsymbol{\eta}^x$. For N_{gp} Gauss integration points, $N \times N_{gp}$ cellular material models are defined concurrently at the microscopic scale.

At the macroscopic scale, topology design variables are defined is association with the element internal force vector $\bar{\mathbf{f}}^i_{\text{int}}$ in analogy to [1.16]

$$\bar{\mathbf{f}}^i_{\text{int}} = \rho_i \int_{\Omega_i} \bar{\mathbf{B}}^T \bar{\boldsymbol{\sigma}} \, \mathrm{d}\Omega_i, \quad i = 1, \ldots, N_{\text{s}} \qquad [4.9]$$

where the effective stress $\bar{\boldsymbol{\sigma}}$ is computed via the volume averaging relation $\bar{\boldsymbol{\sigma}}(\boldsymbol{x}) = \langle \boldsymbol{\sigma}(\boldsymbol{x}, \boldsymbol{y}) \rangle$, where the microscopic stress is computed on the optimized material topology subjected to a prescribed overall strain. In practice, for void elements the optimization at the microscopic scale can be saved and their effective stresses are set directly to zero.

At the microscopic scale, consider the cellular material model corresponding to material point x, the stiffness matrix $\mathbf{K}^x_j$ of the jth element is defined explicitly in terms of η^x_j as

$$\mathbf{K}^x_j = \eta^x_j \mathbf{K}^x_0, \quad j = 1, \ldots, N_x \qquad [4.10]$$

where $\mathbf{K}^x_0$ is the stiffness matrix of the element with full solid material when $\eta_j = 1$.

The simultaneous optimization problem of [4.5] is equivalent to the minimization of the macroscopic compliance f_{c} subjected to macroscopic and microscopic material volume fraction constraints

$$\begin{aligned} \min_{\boldsymbol{\rho}, \boldsymbol{\eta}^x} &: f_{\text{c}} = \bar{\mathbf{f}}^T_{\text{ext}} \bar{\mathbf{u}} \\ \text{subject to} &: \bar{\mathbf{r}}(\bar{\mathbf{u}}, \boldsymbol{\rho}, \boldsymbol{\eta}) = \mathbf{0} \\ &: V_\rho = \sum \rho_i v_i = V^{\text{s}}_{\text{req}} \\ &: V_\eta = \sum \eta_j v^x_j = V^x_{\text{req}} \\ &: \rho_i = 0 \text{ or } 1, \; i = 1, \ldots, N_{\text{s}} \\ &: \eta^x_j = 0 \text{ or } 1, \; j = 1, \ldots, N_x. \end{aligned} \qquad [4.11]$$

where $\bar{\mathbf{f}}_{\text{ext}}^{T}$ is the macroscale external force vector. V_ρ and V_η are the total volumes of solid elements at the two scales, respectively. V_{req}^{s} and V_{req}^{x} are the required volumes of solid elements defined at the two scales, respectively. v_i is the volume of macroscale element i, and v_j^x is the volume of the microscale element j of the material model attached on macro point x. $\bar{\mathbf{u}}$ is the converged macroscale displacement solution. $\bar{\mathbf{r}}(\bar{\mathbf{u}}, \boldsymbol{\rho}, \boldsymbol{\eta}^x)$ stands for the force residual at the macroscopic scale

$$\bar{\mathbf{r}}(\bar{\mathbf{u}}, \boldsymbol{\rho}, \boldsymbol{\eta}^x) = \bar{\mathbf{f}}_{\text{ext}} - \sum_{i=1}^{N} \rho_i \int_{\Omega_i} \bar{\mathbf{B}}^T \bar{\boldsymbol{\sigma}} \mathrm{d}\Omega_i. \qquad [4.12]$$

where Ω_i denotes the region occupied by the ith element.

4.3.2. *Decomposed optimization models*

Equation [4.11] is the discretized model of the optimization problem [4.5]. As presented in section 4.1, the simultaneous design model of [4.5] can be equivalently split into a "master" problem of [4.6] and "slave" problems of [4.7] due to the nature of separability of the design variables $\rho(x)$ and $\eta(x, y)$, which are defined at the two scales, respectively.

The discretized model of the master problem of [4.6] is formulated as

$$\begin{aligned} \min_{\boldsymbol{\rho}} &: f_c(\boldsymbol{\rho}, \bar{\mathbf{u}}) \\ \text{subject to} &: \bar{\mathbf{r}}(\bar{\mathbf{u}}, \boldsymbol{\rho}, \boldsymbol{\eta}) = \mathbf{0} \\ &: V_\rho = \sum \rho_i v_i = V_{\text{req}}^{\text{s}} \\ &: \rho_i = 0 \text{ or } 1, \ i = 1, \ldots, N_{\text{s}} \end{aligned} \qquad [4.13]$$

in terms of $\boldsymbol{\rho}$ defined at the macroscopic structural scale, where the microscale variables $\boldsymbol{\eta}$, which decide structural constitutive behaviors, are implicitly included in the equilibrium equation $\bar{\mathbf{r}}(\bar{\mathbf{u}}, \boldsymbol{\rho}, \boldsymbol{\eta}) = \mathbf{0}$, which is solved by the FE^2-based solution scheme (section 4.2).

For a given value of $\boldsymbol{\rho} = \boldsymbol{\rho}^*$ and the corresponding displacement solution $\bar{\mathbf{u}} = \bar{\mathbf{u}}^*$, with the assumption that cellular material models $\boldsymbol{\eta}^x$ are defined only at material points where $\rho^*(x) = 1$, the discretized models of "slave" problems

[4.6], the microscale material stiffness maximizations subjected to microscale material volume fraction constraints, are defined in the following form

$$\begin{aligned} \max_{\boldsymbol{\eta}^x} &: \bar{w}(\boldsymbol{\eta}^x) \\ \text{subject to} &: \mathbf{K}^x(\boldsymbol{\eta}^x)\mathbf{u}^x = \mathbf{0} \\ &: \langle \boldsymbol{\varepsilon}(\mathbf{u}^x) \rangle = \bar{\boldsymbol{\varepsilon}}(\boldsymbol{x}) \\ &: V_\eta = \sum \eta_j v_j^x = V_{\text{req}}^x \\ &: \eta_j = 0 \text{ or } 1,\ j = 1, \ldots, N_x. \end{aligned} \qquad [4.14]$$

Note that there exists no external force at the microscopic scale. The microscale systems are constrained by means of the imposed periodic boundary conditions, satisfying the equality between $\langle \boldsymbol{\varepsilon}(\mathbf{u}^x) \rangle$ and $\bar{\boldsymbol{\varepsilon}}(\boldsymbol{x})$.

To implement topology optimization, sensitivities of design variables need to be provided. According to [4.9] and [4.13], the sensitivity for the nonlinear design problem at the macroscopic scale is evaluated as [BEN 03]

$$\frac{\partial f_c}{\partial \rho_i} = -\boldsymbol{\lambda}^T \int_{\Omega_i} \bar{\mathbf{B}}^T \bar{\boldsymbol{\sigma}} \mathrm{d}\Omega_i, \qquad [4.15]$$

where $\boldsymbol{\lambda}$ is the adjoint solution of

$$\bar{\mathbf{K}}_{\tan} \boldsymbol{\lambda} = \bar{\mathbf{f}}_{\text{ext}}, \qquad [4.16]$$

where $\bar{\mathbf{K}}_{\tan}$ is the tangent stiffness matrix at the convergence $\bar{\mathbf{u}}_{\text{sol}}$ of the NR solution process. As has been shown in Figure 4.2, the tangent stiffness matrix for a certain displacement solution is the corresponding elastic stiffness matrix itself, i.e. $\bar{\mathbf{K}}_{\tan}(\bar{\mathbf{u}}_{\text{sol}}) = \bar{\mathbf{K}}(\bar{\mathbf{u}}_{\text{sol}})$ and therefore $\boldsymbol{\lambda} = \bar{\mathbf{u}}_{\text{sol}}$. The evaluation of [4.15] can be further simplified to

$$\frac{\partial f_c}{\partial \rho_i} = -\bar{\mathbf{u}}_{\text{sol}}^T \int_{\Omega_i} \bar{\mathbf{B}}^T \bar{\boldsymbol{\sigma}} \mathrm{d}\Omega_i, \qquad [4.17]$$

where Ω_i denotes the region occupied by the ith element.

On the contrary, the design problem at the microscopic scale is self-adjoint [BEN 03] and therefore its sensitivity is evaluated according to [4.10] and [4.14]

$$\frac{\partial \bar{w}(\boldsymbol{\eta}^x)}{\partial \eta_j^x} = \frac{1}{2} (\mathbf{u}_j^x)^T \mathbf{K}_0^x \mathbf{u}_j^x. \qquad [4.18]$$

4.3.3. *BESO updating scheme*

The applied BESO updating scheme for macroscale structural topology optimization exactly follows the scheme presented in section 1.4.3 with the sensitivity numbers defined as

$$\bar{\alpha}_i = -\frac{\partial f_{\mathrm{c}}}{\partial \rho_i}\frac{1}{v_i}. \quad [4.19]$$

The same updating scheme is adopted for microscale material topology optimization except for a minor modification on the filtering scheme. In order to maintain the structural periodicity during the optimization, special attention needs to be paid to the filtering near the boundary region. The filtering domain is enlarged periodically in accordance with the filtering radius $r_{\min}$ as shown in Figure 4.3. The sensitivity numbers for microscale material design is defined as

$$\alpha_j^x = \frac{\partial \bar{w}(\boldsymbol{\eta}^x)}{\partial \eta_j^x}\frac{1}{v_j^x}. \quad [4.20]$$

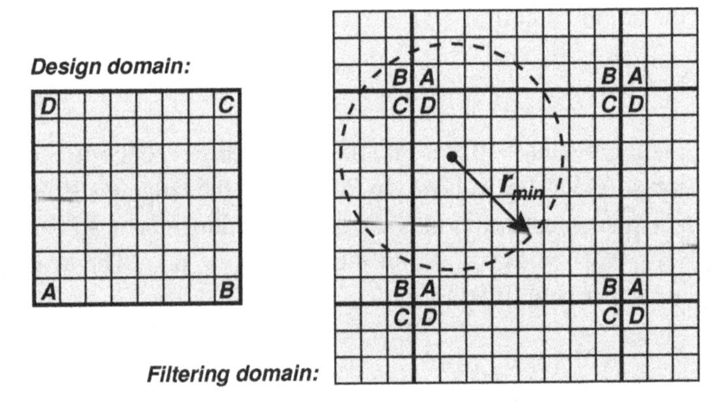

Figure 4.3. *Illustration of periodic filtering scheme*

4.3.4. *Simultaneous design framework*

The simultaneous design framework is outlined in algorithm 4.1. Generally speaking, there exist three layers in this framework according to [4.6]. The very outer layer is the global structural optimization that loops until the design solution $\boldsymbol{\rho}$ reaches convergence. The inner layer is the local cellular material optimization with respect to $\boldsymbol{\eta}^x$ with prescribed overall strain value $\bar{\boldsymbol{\varepsilon}}(\boldsymbol{x})$ at Gauss point $\boldsymbol{x}$. The middle layer ensures compatibility between the two scales, which corresponds to the nonlinear scale-interface equilibrium.

Algorithm 4.1. Simultaneous design framework

1: Initialize $\boldsymbol{\rho}_0$ and $\mathbf{K}_0$;
2: **while** $\|\boldsymbol{\rho}_{i+1} - \boldsymbol{\rho}_i\| > \delta_{\text{opt}}$ $\{i++\}$ **do**
3:     **while** $\|\bar{\mathbf{f}}_{\text{ext}} - \bar{\mathbf{f}}_{\text{int}}\| > \delta_{\text{f}}$ **do**
4:         **loop** over all macro Gauss points
5:             compute the effective strain $\bar{\boldsymbol{\varepsilon}}(\boldsymbol{x})$;
6:             define PBC on the associated material model upon $\bar{\boldsymbol{\varepsilon}}(\boldsymbol{x})$;
7:             initialize microscale material topology $\boldsymbol{\eta}_0^x$;
8:             **while** $\|\boldsymbol{\eta}_{j+1}^x - \boldsymbol{\eta}_j^x\| > \delta_{\text{opt}}^x$ $\{j++\}$ **do**
9:                 compute $\bar{w}^x$ and sensitivities $\partial \bar{w}^x / \partial \boldsymbol{\eta}^x$;
10:                 update $\boldsymbol{\eta}^x$ using the BESO scheme;
11:             **end while**
12:             compute $\bar{\boldsymbol{\sigma}} = \langle \boldsymbol{\sigma} \rangle$ on the optimized material;
13:         **end loop**
14:         NR update: $\bar{\mathbf{K}}_0 \Delta \bar{\mathbf{u}} = \bar{\mathbf{f}}_{\text{ext}} - \sum \rho_i \int_{\Omega_i} \bar{\mathbf{B}}^T \bar{\boldsymbol{\sigma}} \mathrm{d}\Omega_i$;
15:     **end while**
16:     compute f_c and sensitivities $\partial f_c / \partial \boldsymbol{\rho}$;
17:     update $\boldsymbol{\rho}$ using BESO scheme;
18: **end while**
19: **return** $\boldsymbol{\rho}$ and $\boldsymbol{\eta}^x$.

The presented algorithm is defined in a general manner. With minor modifications and additional constraints, the presented algorithm can mutate to various types of design problems. For instance, when topology optimization is limited to a single scale, the presented algorithm corresponds to standard structural topology optimization [HUA 10b] and material microstructure designs [HUA 11]. When an universal microstructure $\boldsymbol{\eta}$ is assumed and designed at the microscopic scale, the algorithm corresponds to the approaches developed in [HUA 13, YAN 14], depending on the definition at the macroscopic scale. When the discrete BESO method is replaced by the continuous SIMP model, the algorithm corresponds to variants of

[ROD 02, ZHA 06], where the nonlinear scale-interface equilibrium is solved by the FE^2 method.

4.4. Numerical examples

In this section, four numerical examples are considered. In section 4.4.1, optimal designs of periodic cellular materials are performed upon different prescribed overall strains. In section 4.4.2, cellular materials are designed for a bridge-type structure, where optimization is performed only at the microscopic scale for the purpose of illustrating the FE^2-based nonlinear computing procedure. In section 4.4.3, simultaneous material and structural design is performed for the same bridge-type structure. Section 4.4.4 further considers the simultaneous design of a half Messerschmitt–Bölkow–Blohm (MBB) beam structure at both macroscopic and microscopic scales.

4.4.1. *Cellular material designs*

Material microstructure designs are considered in this example as the inner problem of [4.6]. The square RVE of cellular material is discretized into 80×80 4-node bilinear elements and $M = 80 \times 80$ density design variables are correspondingly defined. Young's modulus and Poisson's ratio of solid material are set to 1 and 0.3, respectively. The volume constraint of solid material is 60%. The evolution rate in BESO method is set to $c_{\text{er}} = 0.02$. In order to obtain the so called one-length scale microstructure [BEN 03], i.e. avoid too detailed microstructures inside the cells, the filter radius is set to $r_{\min} = 12$.

Due to the applied periodic boundary conditions, an initial guess design has to be defined to trigger topological changes. Here, four soft elements are assigned at the center of the design domain [HUA 11], as shown in Figure 4.4. We consider three representative loading cases where the macroscopic strains are: $(1, 0, 0)^T$, $(0, 0, 1)^T$ and $(1.2, 0.8, 0.5)^T$. The design evolutions and results of the three test cases are shown in Figures 4.4(a)–(c), correspondingly. It can be seen that the design results for horizontal stiffness maximization (Figure 4.4(a)) and shear modulus maximization (Figure 4.4(b)) are quite similar to the design results in [HUA 11, NEV 00, ZHA 07]. Moreover, the spatial periodicity is also guaranteed due to the imposed periodic boundary conditions and periodic filtering scheme, as can be seen from all three test cases.

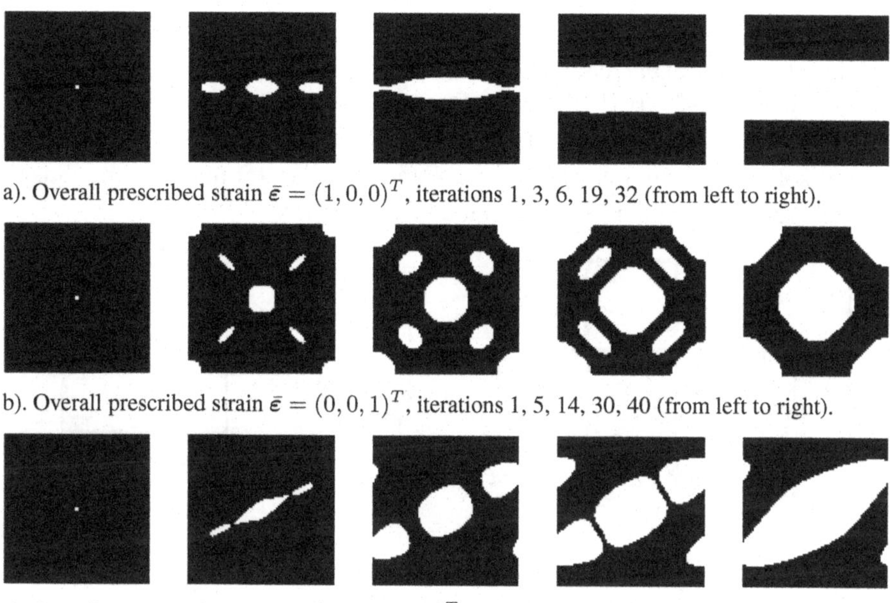

a). Overall prescribed strain $\bar{\varepsilon} = (1,0,0)^T$, iterations 1, 3, 6, 19, 32 (from left to right).

b). Overall prescribed strain $\bar{\varepsilon} = (0,0,1)^T$, iterations 1, 5, 14, 30, 40 (from left to right).

c). Overall prescribed strain $\bar{\varepsilon} = (1.2,0.8,0.5)^T$, iterations 2, 12, 22, 42 (from left to right).

Figure 4.4. *The topological evolutions of the optimal design of cellular materials*

4.4.2. *Material design of a bridge-type structure*

For the purpose of illustrating the FE^2-based nonlinear computing procedure, a simple bridge-type structure as shown in Figure 4.5 is considered, where optimization design is performed only at the inner layer of the problem in [4.6]. The bridge-type structure is discretized into quadrilateral 8-node elements. From the reduced integration scheme, four Gauss integration points are defined for each finite element. Each integration point is attributed with a cellular material model discretized into 80×80 4-node bilinear elements, which means in total $N_{\text{gp}} = 4 \times 8$, 32 cellular material models are considered concurrently at the microscopic scale. Young's modulus and Poisson's ratio of solid material at the microscopic scale are set to be 1 and 0.3, respectively. The volume constraint for each cellular material model is set to 60%. The evolution rate and filter radius in the BESO method are set to $c_{\text{er}} = 0.02$ and $r_{\text{min}} = 12$, the same as defined in the first example.

The evolution of cellular materials in reaching the macroscopic equilibrium is shown in Figure 4.6. The initial macroscopic structural stiffness matrix $\mathbf{K}_0$ with solid materials is used to perform the iterative resolution. The displacement convergence criterion is $\delta = \|\mathbf{u}^{(k+1)} - \mathbf{u}^{(k)}\|_2 / \|\mathbf{u}^{(k)}\|_2 \leq 10^{-2}$. Figure 4.6(a) can be viewed as a more detailed design solution of the composite laminate orientation design [GAO 12]. The design solution varies iteratively from Figures 4.6(a)–(f) upon the nonlinear iterative resolution scheme as presented in section 4.2.

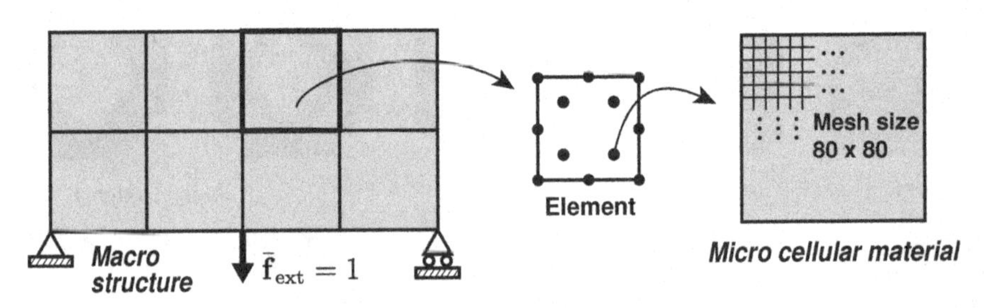

Figure 4.5. *Illustration of the bridge-type structure for cellular material design*

The difference between Figures 4.6(a) and (f) demonstrates the necessity of considering the nonlinearity of the middle layer macro–micro interface problem of [4.6]. Note that Figure 4.6 is a zoomed view of the design results, where the optimized cellular materials corresponding to the Gauss points are enlarged for the purpose of illustration. Upon the homogenization theory, the optimized cellular material only represents the optimal solution at the microscopic scale for that material point, i.e. Gauss integration point. Therefore, the optimized cellular materials represent only the optimal solutions at the associated Gauss points while they are not necessarily contiguous with each other.

4.4.3. *Simultaneous design of a bridge-type structure*

Since this example, optimal designs are performed simultaneously at both macroscopic and microscopic scales. The same bridge-type structure of Figure 4.5 is considered again here in Figure 4.7 while with much finer discretization in order to describe the structural topological changes. The macroscopic structure is discretized into 40×20 4-node bilinear elements, which means in total $N_{\text{gp}} = 4 \times 40 \times 20$, 3200 cellular material models are considered concurrently at the microscopic scale. At the macroscopic scale, $N = 40 \times 20$, 800 density design variables are accordingly defined. For the purpose of limiting the computational requirement, microscopic cellular material model is discretized into 40×40 4-node bilinear elements with $M = 40 \times 40$ design variables. There exist in total $N + N_{\text{gp}} \times M = 5,120,800$ topology design variables in this design problem. At the structural scale, volume constraint of solid material is set to 60%. The other related parameters in the BESO method are evolution rate $c_{\text{er}} = 0.02$ and filter radius $r_{\min} = 3$. At the microscopic material scale, volume constraints are also set to 60%, the evolution rate and filter radius in the BESO method are $c_{\text{er}} = 0.02$ and $r_{\min} = 6$.

(a) substep 1, $\delta = 0.487$, $c^{(1)} = 3.39$

(d) substep 4, $\delta = 0.034$, $c^{(4)} = 10.98$

(b) substep 2, $\delta = 0.217$, $c^{(2)} = 7.29$

(e) substep 5, $\delta = 0.017$, $c^{(5)} = 11.73$

(c) substep 3, $\delta = 0.078$, $c^{(3)} = 9.55$

(f) substep 6, $\delta = 0.009$, $c^{(6)} = 12.13$

Figure 4.6. *Evolution of cellular materials reaching the macroscopic equilibrium*

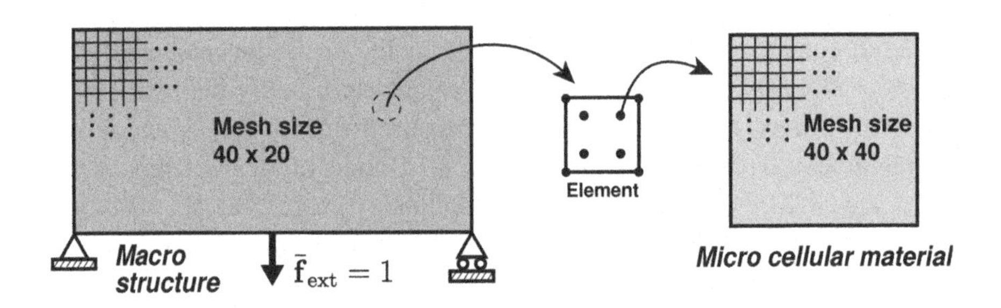

Figure 4.7. *Illustration of the bridge-type structure for simultaneous design*

For the purpose of illustrating the nonlinearity due to the adaptation of microstructures, a comparison between the first substep and the converged solutions in the first design iteration is shown in Figure 4.8. The same displacement convergence criterion is used as in the previous example. For each structural design iteration, it takes seven substeps to reach the macroscopic equilibrium. An obvious difference between the two solutions can be observed. The topological evolution of the macroscopic structure together with the converged cellular materials is shown in Figure 4.9. The convergence histories of the compliance and of the volume ratio at the macroscopic scale are demonstrated in Figures 4.10(a) and (b). It takes several more iterations to converge where however invisible differences can be found among the topology results of the last iterations. For this reason, we only kept the historical information before the design iterations with the converged topologies. Similar treatment is used in the following example. The standard monoscale solution obtained using the BESO method with linear elasticity is given in Figure 4.11(b) for the purpose of comparison with Figure 4.11(a). Some typical microstructures obtained in the nonlinear two-scale design is also given in Figure 4.11(c).

Similar to the previous example, the optimized cellular material only represents the optimal solution at the microscopic scale for that material point satisfying the assumptions of scale separation and periodicity. The optimized cellular materials represent only the optimal solutions at associated Gauss points while they are not necessarily contiguous. For manufacturing consideration, denser material points need to be considered at the macroscopic scale and then image-based interpolation schemes such as developed in [XIA 13a] can be employed for generating intermediate microstructures.

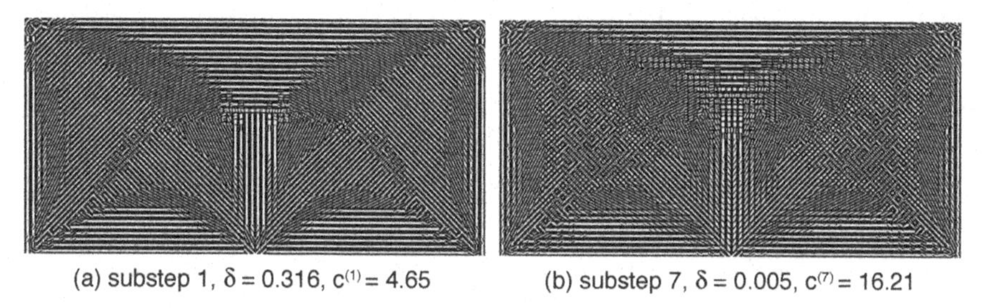

(a) substep 1, $\delta = 0.316$, $c^{(1)} = 4.65$ (b) substep 7, $\delta = 0.005$, $c^{(7)} = 16.21$

Figure 4.8. *The initial and converged design solutions of the first design iteration*

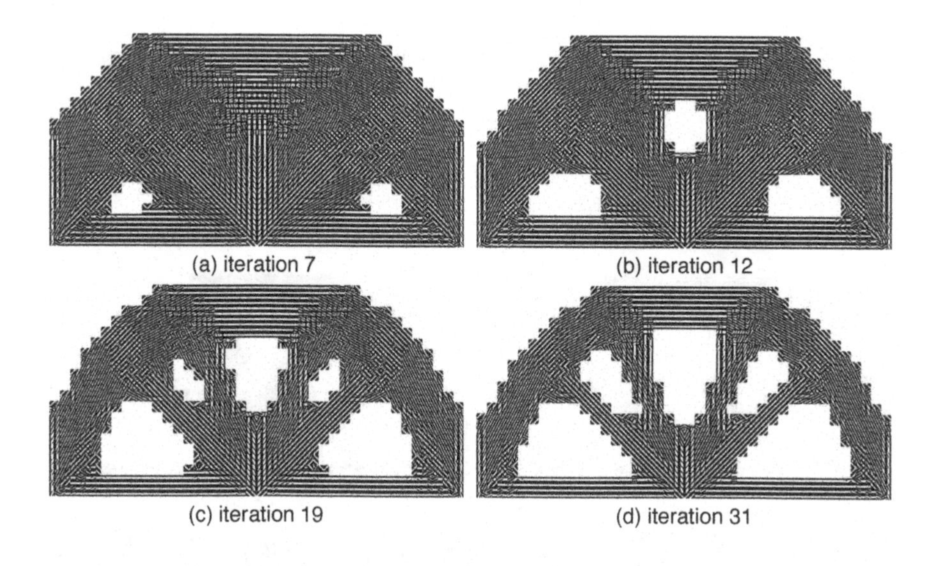

(a) iteration 7 (b) iteration 12

(c) iteration 19 (d) iteration 31

Figure 4.9. *Evolution of cellular materials and structure during the simultaneous designs*

Each individual macroscopic design iteration can also be viewed as an independent design problem with fixed topology and with varying microstructures (section 4.4.2); therefore, the illustrated cellular material topologies shown in Figures 4.9(a)–(f) can be viewed as the optimal design solutions for the six macroscopic structures. Similarly, the final simultaneous optimal design (Figure 4.9(f)) can also be interpreted as the general case of the simultaneous structural topology and laminate orientation designs [SET 05, GAO 13], where the orientation variables are defined in an elementwise manner. In this work, anisotropic cellular materials are defined in a pointwise manner, which of course leads to more intensive computational requirement. As can be observed in Figure 4.9(f), uniaxial materials may be sufficient at the main branches of the structure; while in order to have a higher structural performance, anisotropic materials have to be used at the joints of the main branches due to the more complex internal forces.

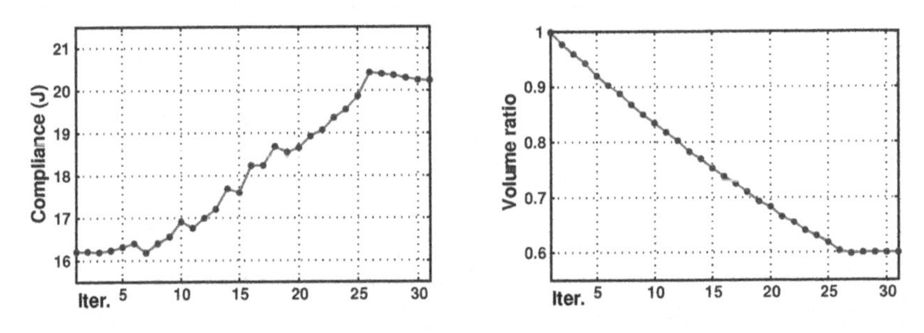

Figure 4.10. *Convergence histories of compliance and volume ratio at the macroscopic scale*

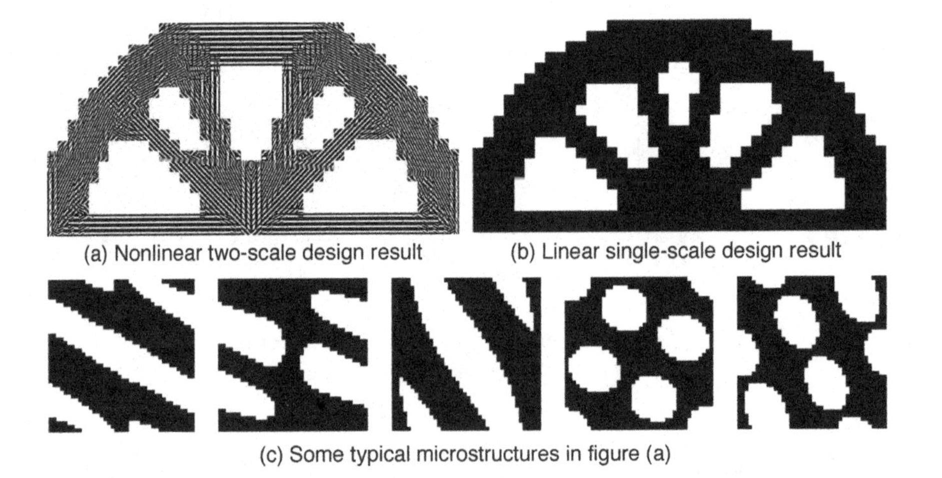

Figure 4.11. *Comparison between the nonlinear two-scale and linear monoscale solutions*

4.4.4. *Simultaneous design of a half MBB beam*

The so-called MBB beam [BEN 03] is further considered in this example. The same example has also been investigated in [ROD 02] for the simultaneous design using the continuous SIMP model, and recently in [GAO 12] for composite laminate orientation design. Due to the symmetry of the problem, only a half MBB beam is considered, as shown in Figure 4.12. The macroscopic structure is discretized into 40×16 4-node bilinear elements, which means in total $N_{\text{gp}} = 4 \times 40 \times 16 = 2560$ cellular material models are considered concurrently at the microscopic scale. At the macroscopic structure, $N = 40 \times 16 = 640$ density design variables are accordingly defined. The microscopic cellular material model is discretized into 40×40 4-node bilinear elements with $M = 40 \times 40$ design variables. There exist in total $N + N_{\text{gp}} \times M = 4,096,640$ topology design variables in this design problem. Volume constraints of solid material are set to 60% at both scales. The other parameters for the BESO method at the both scales are the same as defined in the previous example.

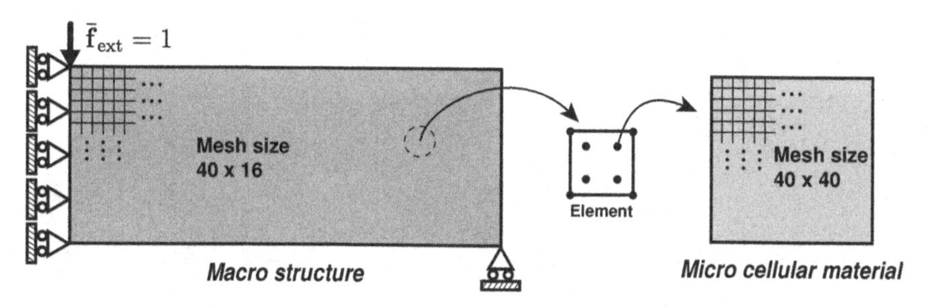

Figure 4.12. *Illustration of the half MBB beam problem for simultaneous design*

Similarly, a comparison between the first substep and the converged solutions in the first design iteration is given in Figure 4.13. The same displacement convergence criterion is used as in the previous examples. In this example, it takes around six substeps to reach the macroscopic equilibrium for each design iteration. The simultaneous evolutions of both cellular materials and structure are shown in Figure 4.14. The convergence histories of the compliance and the volume ratio at the macroscopic scale are demonstrated in Figures 4.15(a) and (b), respectively. The standard monoscale solution obtained using BESO with linear elasticity is given in Figure 4.16(b) for the purpose of comparison with Figure 4.16(a). Some typical microstructures obtained in the nonlinear two-scale design are also given in Figure 4.16(c). Similar to previous examples, the optimized cellular material only represents the optimal solution at the microscopic scale for that material point satisfying the assumptions of scale separation and periodicity. The optimized cellular materials represents only the optimal solutions at the associated Gauss points while are not necessarily contiguous with each other.

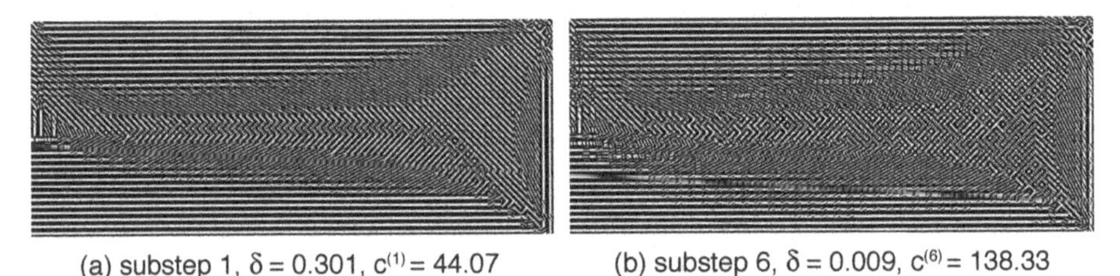

Figure 4.13. *The initial and converged design solutions of the first design iteration*

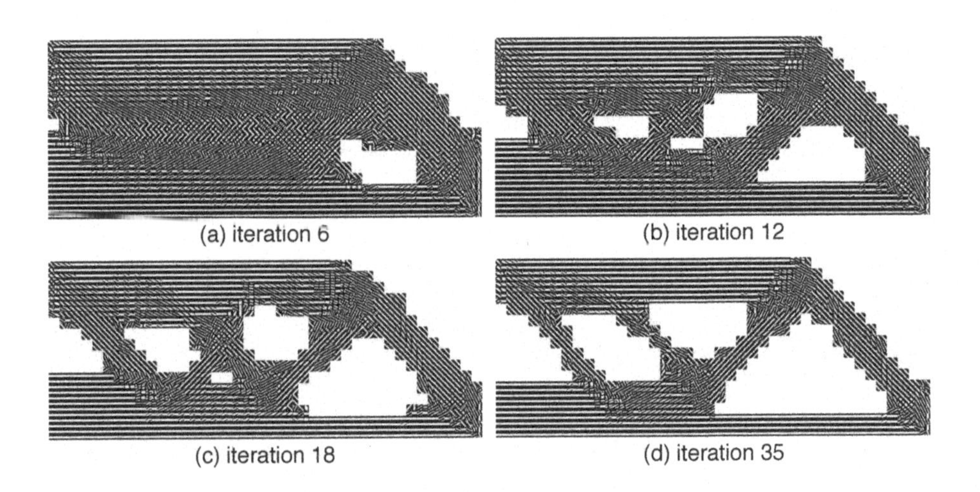

Figure 4.14. *Evolution of cellular materials and structure during the simultaneous designs*

Though the macroscopic structural topological evolution in Figure 4.14 is quite similar to conventional monoscale structural topology design result, it is however the result of considering the optimized cellular materials. Each individual design result in Figure 4.14 is in fact an optimal design solution for the considered macroscopic structure. From Figure 4.14(f), we can see that uniaxial materials are obtained along the main branches, while anisotropic materials are obtained at the joints of the main branches.

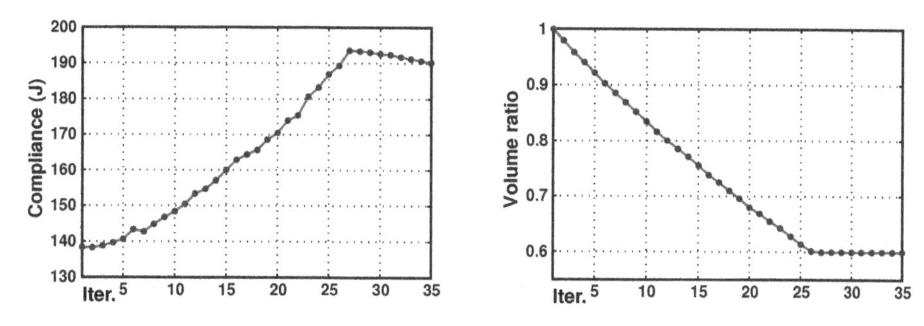

Figure 4.15. *Convergence histories of compliance and volume ratio at the macroscopic scale*

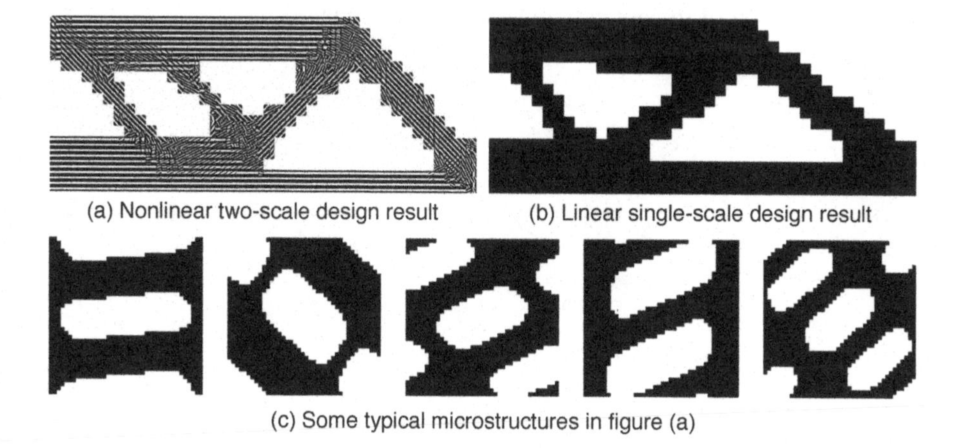

Figure 4.16. *Comparison between the nonlinear two-scale and linear monoscale solutions*

4.5. Concluding remarks

In this chapter, we have extended the multiscale design framework for simultaneous topology optimization of structure and material microstructures.

The proposed model enables us to obtain optimal structures with spatially varying properties realized by the simultaneous design of microstructures. Note that the designed structure with varying constitutive behaviors due to the microstructures are constituted in fact by only one base material, which greatly favors the process of 3D printing that a single material can usually be used for fabrication.

The nonlinear macro–micro interface equilibrium is solved using the initial stiffness NR solution scheme at the expense of the convergence rate. A more efficient iterative solution scheme needs to be constructed for higher computing efficiency, which is still an open issue for the moment.

The simultaneous design results in an intensive computational cost even for the linear case. Currently, we have treated the optimization process of material microstructure as a generalized constitutive behavior, which means that the design efficiency can be improved drastically by a straightforward application of the existing ROMs for nonlinear materials. In the following chapter, we directly apply the reduced database model NEXP [YVO 09] for the representation of the generalized constitutive behavior.

5

Reduced Database Model for Material Microstructure Optimizations

This chapter builds upon Chapter 4 in which simultaneous topology optimization has been performed for both structure and the underlying materials. As has been discussed in Chapter 4, the simultaneous design framework requires intensive computational cost due to large number of repetitive local material microstructure optimizations. Note that one particular feature of simultaneous design strategy is treating the optimization process of material microstructure as a generalized nonlinear constitutive behavior. With this feature in mind, this chapter is dedicated to improving the simultaneous design efficiency by a straightforward application of the NEXP model [YVO 09] to approximate the material microstructure behavior.

The NEXP model was initially developed with the intent of representing the effective behaviors of nonlinear elastic composites at small strains and extended later for finite strains [YVO 13]. From the NEXP model, a database is built from a set of numerical experiments of local material optimizations in the space of macroscale strain. Each value in the database corresponds to the strain energy density evaluated on a material microstructure, optimized according to the imposed macroscopic strain. From tensor decomposition, a continuous representation of the strain energy density is built as a sum of products of one-dimensional interpolation functions. As a result of this off-line step, the effective strain–energy and stress–strain relations required for macroscale structural evaluation and optimization are provided in a numerically explicit manner. The results given by the reduced database model are compared with full-scale results. The explicit material behavior representation given by the NEXP model is then used to serve the simultaneous design at a negligible computational cost.

The remainder of this chapter is organized as follows: Section 5.1 briefly reviews the simultaneous design framework. Section 5.2 gives the generalized constitutive behavior of locally optimized materials. Section 5.3 presents the construction strategy of the reduced database model. Section 5.4 gives the structural topology optimization model using the constitutive represented by the reduced database model. Section 5.5 gives the general design algorithm. Section 5.6 showcases the developed model by means of numerical test examples. Concluding remarks are given in section 5.7

5.1. Simultaneous design framework

In section 4.1, we briefly reviewed the simultaneous design framework. In analogy to [4.1], the general formulation for the minimum compliance problem is [BEN 03]

$$\max_{(\rho,\eta)\in\mathcal{A}_{\mathrm{ad}}} \min_{u\in\mathcal{U}} \left\{ \frac{1}{2} \int_{\Omega} C_{ijkh}\left(x, \rho(x), \eta(x,y)\right) \frac{\partial u_i}{\partial x_j} \frac{\partial u_k}{\partial x_h} \mathrm{d}\Omega - l(u) \right\}. \qquad [5.1]$$

Following [BEN 95, THE 99], the separation of the two scale variables and the interchange of the equilibrium and local optimizations of [5.1] result in a reformulated problem as [4.6]

$$\max_{\rho\in\mathcal{A}_\rho} \min_{u\in\mathcal{U}} \left\{ \int_{\Omega} \bar{w}\left(x, u, \eta(x,y)\right) \mathrm{d}\Omega - l(u) \right\}, \qquad [5.2]$$

where the pointwise maximization of the strain energy density $\bar{w}$

$$\bar{w} = \max_{\eta\in\mathcal{A}_\eta} \frac{1}{2} \rho(x) C_{ijkh}\left(x, \eta(x,y)\right) \frac{\partial u_i}{\partial x_j} \frac{\partial u_k}{\partial x_h} \qquad [5.3]$$

is treated as a subproblem defined in analogy to [4.7].

In the case of discrete topology design models (e.g. BESO, [YAN 14, XIA 14a]), $\mathcal{A}_\rho$ and $\mathcal{A}_\eta$ are simply defined as

$$\mathcal{A}_\rho = \left\{ \rho \mid \rho = 0 \text{ or } 1, \int_\Omega \rho(x) \mathrm{d}\Omega = V_{\text{req}}^{\text{s}} \right\}, \quad [5.4]$$

$$\mathcal{A}_\eta = \left\{ \eta \mid \eta = 0 \text{ or } 1, \int_{\Omega_x} \eta(x,y) \mathrm{d}\Omega^x = V_{\text{req}}^x \right\}, \quad [5.5]$$

in analogy to [4.3] and [4.4], where $V_{\text{req}}^{\text{s}}$ and V_{req}^x are the allowed material volume at the macro- and microscales, respectively. Note that V_{req}^x can vary from point to point.

Note that, because cellular materials at the microscopic scale are optimized in response to the current strain loading statuses (Figure 5.1) while the optimized cellular materials in turn update the macroscopic constitutive behavior, the scale-interface equilibrium is therefore in general nonlinear even when linear models are assumed at both scales. To solve the nonlinear scale-interface equilibrium problem, the FE^2 method is employed to bridge the two separated scales with the initial stiffness NR solution scheme (section 4.2).

5.2. Generalized material constitutive behavior

The simultaneous design framework [5.2] requires solving at the microscopic scale the material stiffness maximization problem [5.3]

$$\begin{aligned} \max_{\eta} &: \bar{w}(x, \bar{\varepsilon}) \\ \text{subject to} &: \operatorname{div} \boldsymbol{\sigma}(x,y) = 0 \\ &: \langle \boldsymbol{\varepsilon}(x,y) \rangle = \bar{\varepsilon}(x) \\ &: V(\eta) = \textstyle\int_{\Omega^x} \eta \mathrm{d}\Omega^x = V_{\text{req}}^x \\ &: \eta(x,y) = 0 \text{ or } 1, y \in \Omega^x \end{aligned} \quad [5.6]$$

for each given value of $\bar{\varepsilon}(x)$, imposed as PBC on the cellular material model. Upon the assumption of periodicity, the microscopic displacement field is a sum of a macroscopic displacement field and a periodic fluctuation field $\boldsymbol{u}^*$, whose volume average over Ω^x equals zero ($\langle \boldsymbol{\varepsilon}(\boldsymbol{u}^*) \rangle = 0$) leading to $\langle \boldsymbol{\varepsilon}(x,y) \rangle = \bar{\varepsilon}(x)$ [MIC 99].

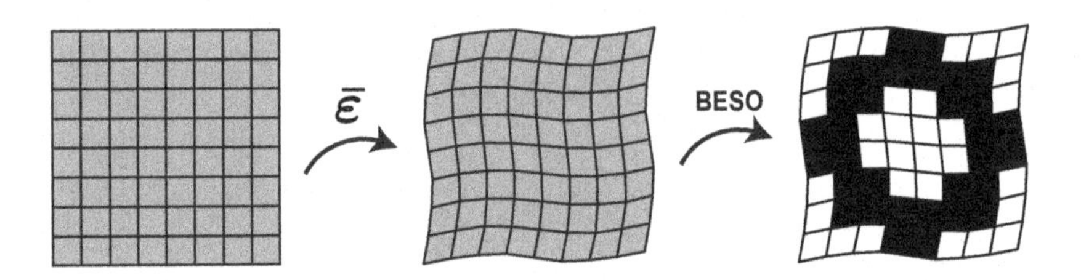

Figure 5.1. *Illustration of the microscopic material optimization procedure*

For an optimized cellular material with a specified topology $\eta = \eta^*$ as shown in Figure 5.1, we have the following effective stress–strain relationship at the macroscopic scale

$$\bar{\boldsymbol{\sigma}} = \bar{\mathbb{C}}^{\mathrm{hom}}(\eta^*) : \bar{\boldsymbol{\varepsilon}}, \quad [5.7]$$

where $\bar{\mathbb{C}}^{\mathrm{hom}}(\eta^*)$ is the homogenized elastic tensor, which can be determined by solving the microscale boundary value problem for six independent overall strain values in general 3D case. Note that, with chosen optimization algorithm and volume constraint V^x_{req}, the optimized value η^* in [5.6] is dependent only on the imposed overall macroscale strain tensor value $\bar{\boldsymbol{\varepsilon}}$, and thus we can restate [5.7] as

$$\bar{\boldsymbol{\sigma}} = \bar{\mathbb{C}}^{\mathrm{hom}}(\bar{\boldsymbol{\varepsilon}}) : \bar{\boldsymbol{\varepsilon}}, \quad [5.8]$$

where $\bar{\mathbb{C}}^{\mathrm{hom}}(\bar{\boldsymbol{\varepsilon}})$ depends on $\bar{\boldsymbol{\varepsilon}}$ through solving [5.6] in a nonlinear manner.

Let $\bar{w}(x, \bar{\boldsymbol{\varepsilon}})$ denote the macroscale strain energy density

$$\bar{w}(x, \bar{\boldsymbol{\varepsilon}}) = \frac{1}{2}\langle \boldsymbol{\sigma} : \boldsymbol{\varepsilon} \rangle = \frac{1}{2|\Omega_x|} \int_{\Omega_x} \boldsymbol{\sigma} : \boldsymbol{\varepsilon} \, \mathrm{d}\Omega_x \quad [5.9]$$

at macroscopic point x of the optimized local material with strain $\bar{\boldsymbol{\varepsilon}}$. Based on Hill's energy condition [HIL 52], we have

$$\langle \boldsymbol{\sigma} : \boldsymbol{e} \rangle = \bar{\boldsymbol{\sigma}} : \bar{\boldsymbol{\varepsilon}}. \quad [5.10]$$

It is worth noting that for the strain-based formulation, the equilibrium problem [5.2] remains convex after the local material optimizations [BEN 03]. The effective stress–strain relationship can be derived as [HIL 63]

$$\bar{\sigma} = \frac{\partial \bar{w}(x, \bar{\varepsilon})}{\partial \bar{\varepsilon}}. \quad [5.11]$$

Note that, when the microscale optimization problem settings are identical for all macroscale points, i.e. V_{req}^x is set to a constant value, which is the case in the following context, the variable x denoting the associated macroscale point in $\bar{w}(x, \bar{\varepsilon})$ can therefore be omitted. In viewing the local material optimization process of [5.6] as a particular regime of material nonlinearity, the main objective of the present work is to construct an explicit representation of $\bar{w}(\bar{\varepsilon})$ over the tensor space $\mathscr{E}$ spanned by $\bar{\varepsilon}$ such that the simultaneous design can be performed with an effective stress–strain relationship provided at an extremely reduced computational cost. Note that there is no closed-form expression of $\bar{w}(\bar{\varepsilon})$ over $\mathscr{E}$ when the local material is constituted by multiple physically or geometrically nonlinear material phases. For such reason, we choose to follow the NEXP strategy [YVO 09] to construct an approximate expression of $\bar{w}(\bar{\varepsilon})$ using a precomputed database.

5.3. Reduced database model (NEXP)

In the following, we make a step further and extend the NEXP model to represent the new regime of nonlinearity due to locally optimized materials.

5.3.1. *Basic reduction strategy*

The NEXP model aims to construct an explicit approximation or response surface $\bar{w}(\bar{\varepsilon})$ over $\mathscr{E}$ using a precomputed database and interpolation schemes (see Figure 5.2), with an expectation that $\tilde{w}(\bar{\varepsilon})$ approaches enough to $\bar{w}(\bar{\varepsilon})$

$$\bar{w}(\bar{\varepsilon}) \approx \tilde{w}(\bar{\varepsilon}) = \sum_q N_q(\bar{\varepsilon})\bar{w}_q, \quad [5.12]$$

where N_q are interpolation functions and $\bar{w}_q$ are the strain energy density values stored in the database, which are evaluated by means of a set of numerical experiments over a test space $\mathscr{E}$. It is important to emphasize that

$\bar{w}_q$ used to construct the response surface as indicated in Figure 5.2 corresponds to the energy density of a locally optimized material for a given test value $\bar{\varepsilon}_q$. Once the database model is built, the effective stress–strain relationship and tangent elastic stiffness tensor $\bar{\mathbb{C}}^{\tan}$ can be explicitly obtained as

$$\bar{\boldsymbol{\sigma}} \approx \sum_q \frac{\partial N_q(\bar{\varepsilon})}{\partial \bar{\varepsilon}} \bar{w}_q, \qquad [5.13]$$

and

$$\bar{\mathbb{C}}^{\tan} \approx \sum_q \frac{\partial^2 N_q(\bar{\varepsilon})}{\partial \bar{\varepsilon} \partial \bar{\varepsilon}} \bar{w}_q, \qquad [5.14]$$

provided the interpolation functions N_q are at least twice continuously differentiable.

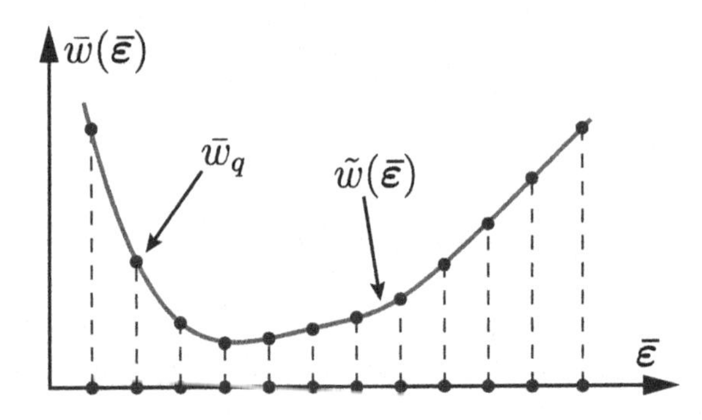

Figure 5.2. *Interpolated strain energy density function $\tilde{w}(\bar{\varepsilon})$ from values $\bar{w}_q$*

5.3.2. *Notations and test space*

We employ Voigt notation and the average stress tensor $\bar{\boldsymbol{\sigma}}$ and strain tensor $\bar{\varepsilon}$ are represented by six-dimensional vectors

$$\begin{cases} \bar{\boldsymbol{\sigma}} = (\bar{\sigma}_1, \bar{\sigma}_2, \bar{\sigma}_3, \bar{\sigma}_4, \bar{\sigma}_5, \bar{\sigma}_6)^T \equiv (\bar{\sigma}_{11}, \bar{\sigma}_{22}, \bar{\sigma}_{33}, \bar{\sigma}_{23}, \bar{\sigma}_{13}, \bar{\sigma}_{12})^T, \\ \bar{\varepsilon} = (\bar{\varepsilon}_1, \bar{\varepsilon}_2, \bar{\varepsilon}_3, \bar{\varepsilon}_4, \bar{\varepsilon}_5, \bar{\varepsilon}_6)^T \equiv (\bar{\varepsilon}_{11}, \bar{\varepsilon}_{22}, \bar{\varepsilon}_{33}, 2\bar{\varepsilon}_{23}, 2\bar{\varepsilon}_{13}, 2\bar{\varepsilon}_{12})^T. \end{cases} \qquad [5.15]$$

In the 3D case, the strain energy density function for one macroscopic point x can now be expressed over a six-dimensional vector space

$$\bar{w}(\bar{\varepsilon}) = \bar{w}(\bar{\varepsilon}_1, \bar{\varepsilon}_2, \ldots, \bar{\varepsilon}_6) \quad [5.16]$$

and individual components of [5.13] and [5.14] are given by

$$\bar{\sigma}_i \approx \sum_q \frac{\partial N_q(\bar{\varepsilon})}{\partial \bar{\varepsilon}_i} \bar{w}_q, \quad \bar{C}_{ij}^{\tan} \approx \sum_q \frac{\partial^2 N_q(\bar{\varepsilon})}{\partial \bar{\varepsilon}_i \partial \bar{\varepsilon}_j} \bar{w}_q. \quad [5.17]$$

Following [YVO 09], we discretize the test space $\mathscr{E}$ into a regular field $\Delta = \Delta_1 \times \Delta_2 \times \cdots \times \Delta_D$, where Δ_i are one-dimensional domains corresponding to the components $\bar{\varepsilon}_i$ and $D = 3$ or 6 for 2D and 3D problems, respectively. Each dimension is discretized into p nodes and in total we have p^D nodes in the database. Because linear material models are assumed at both scales, the determination of the interval of Δ_i is straightforward. All components of $\bar{\varepsilon}$ can be scaled between -1 and 1 by dividing the maximum component term

$$\hat{\varepsilon} = \frac{1}{\bar{\varepsilon}_{\max}} \bar{\varepsilon} \in [-1, 1] \times \cdots \times [-1, 1], \quad [5.18]$$

where

$$\bar{\varepsilon}_{\max} = \max\left(|\bar{\varepsilon}_1|, |\bar{\varepsilon}_2|, \ldots, |\bar{\varepsilon}_6|\right). \quad [5.19]$$

Correspondingly, we have

$$\bar{w}(\bar{\varepsilon}) = \bar{\varepsilon}_{\max}^2 \bar{w}(\hat{\varepsilon}), \quad \bar{\boldsymbol{\sigma}} = \bar{\varepsilon}_{\max} \frac{\partial \bar{w}(\hat{\varepsilon})}{\partial \hat{\varepsilon}}, \quad \bar{\mathbb{C}}^{\tan} = \frac{\partial \bar{w}^2(\hat{\varepsilon})}{\partial \hat{\varepsilon} \partial \hat{\varepsilon}} \quad [5.20]$$

in which $\bar{w}(\hat{\varepsilon})$ is the strain energy density evaluated using the scaled effective strain $\hat{\varepsilon}$.

It is worth noting that if the considered material optimization problem possesses symmetries, then many computations can be avoided. Moreover, the adopted space discretization strategy can be further optimized or improved, which is of essential importance in saving computational cost for the 3D case.

5.3.3. *Separation of variables and interpolation*

Still following [YVO 09], the precomputed full database is further approximated by a sum of products of one-dimensional interpolation functions via higher order tensor decomposition. The tensor decomposed database requires only one-dimensional interpolations for effective stress and tangent matrices evaluations, which further reduces computing time.

Let $\mathbb{W}$ denote the hypermatrix, which stores the database. It can be approximated in a tensor-decomposed representation [KIE 10]

$$\mathbb{W} \approx \sum_{r=1}^{R} \phi_1^r \otimes \phi_2^r \otimes \cdots \otimes \phi_6^r, \qquad [5.21]$$

where ϕ_j^r are real-valued vectors corresponding to the effective strain tensor components $\bar{\varepsilon}_j$ and R is the number of expanded terms. The vectors ϕ_j^r involved in [5.21] are determined by solving the following least square problem for a given value of R

$$\inf_{\phi_j^r} \left\| \mathbb{W} - \sum_{r=1}^{R} \phi_1^r \otimes \phi_2^r \otimes \cdots \otimes \phi_6^r \right\|^2, \qquad [5.22]$$

where $r = 1, \ldots, R$, $j = 1, \ldots, 6$ and $\|\cdot\|$ is the Frobenius norm. Iterative solution such as alternated least squares is required to solve this nonlinear minimization problem [CAR 70, ZHA 02]. In this study, tensor decomposition of $\mathbb{W}$ is determined using the study of Bader and Kolda [BAD 10].

Once the decomposed vectors in [5.21] are obtained, the continuous representation of $\bar{w}(\bar{\varepsilon})$ written in terms of separated components can be constructed by interpolation

$$\bar{w}(\bar{\varepsilon}_1, \bar{\varepsilon}_2, \ldots, \bar{\varepsilon}_6) \approx \sum_{r=1}^{R} \tilde{\phi}_1^r(\bar{\varepsilon}_1)\tilde{\phi}_2^r(\bar{\varepsilon}_2) \cdots \tilde{\phi}_6^r(\bar{\varepsilon}_6), \qquad [5.23]$$

where $\tilde{\phi}_j^r(\bar{\varepsilon}_j)$ are the interpolated values of ϕ_j^r

$$\tilde{\phi}_j^r(\bar{\varepsilon}_j) = \sum_{q=1}^{Q} N_q(\bar{\varepsilon}_j)\{\phi_j^r\}_q, \qquad [5.24]$$

in which N_q is the one-dimensional $\mathscr{C}^2$ interpolation function associated with the node q; Q denotes the number of nodes supporting the shape functions $N_q(\bar{\varepsilon}_j)$ whose values at $\bar{\varepsilon}_j$ are different from zero. With [5.23], the effective stress can be evaluated by

$$\bar{\sigma}_j(\bar{\varepsilon}_1, \bar{\varepsilon}_2, \ldots, \bar{\varepsilon}_6) \approx \sum_{r=1}^{R} \left(\left\{ \prod_{k \neq j} \tilde{\phi}_k^r(\bar{\varepsilon}_k) \right\} \frac{\partial \tilde{\phi}_j^r(\bar{\varepsilon}_j)}{\partial \bar{\varepsilon}_j} \right), \qquad [5.25]$$

with

$$\frac{\partial \tilde{\phi}_j^r(\bar{\varepsilon}_j)}{\partial \bar{\varepsilon}_j} = \sum_{q=1}^{Q} \frac{\partial N_q(\bar{\varepsilon}_j)}{\partial \bar{\varepsilon}_j} \{\phi_j^r\}_q. \qquad [5.26]$$

Similarly, we have effective stiffness matrix reformulated as

$$C_{ij}^{\tan}(\bar{\varepsilon}_1, \bar{\varepsilon}_2, \ldots, \bar{\varepsilon}_6) \approx \sum_{r=1}^{R} \left(\left\{ \prod_{k \neq i,j} \tilde{\phi}_k^r(\bar{\varepsilon}_k) \right\} \frac{\partial \tilde{\phi}_i^r(\bar{\varepsilon}_i)}{\partial \bar{\varepsilon}_i} \frac{\partial \tilde{\phi}_j^r(\bar{\varepsilon}_j)}{\partial \bar{\varepsilon}_j} \right) \quad \text{if } i \neq j, \qquad [5.27]$$

and

$$C_{ij}^{\tan}(\bar{\varepsilon}_1, \bar{\varepsilon}_2, \ldots, \bar{\varepsilon}_6) \approx \sum_{r=1}^{R} \left(\left\{ \prod_{k \neq j} \tilde{\phi}_k^r(\bar{\varepsilon}_k) \right\} \frac{\partial^2 \tilde{\phi}_j^r(\bar{\varepsilon}_j)}{\partial \bar{\varepsilon}_j^2} \right) \quad \text{if } i = j, \qquad [5.28]$$

with

$$\frac{\partial^2 \tilde{\phi}_j^r(\bar{\varepsilon}_j)}{\partial \bar{\varepsilon}_j^2} = \sum_{q=1}^{Q} \frac{\partial^2 N_q(\bar{\varepsilon}_j)}{\partial \bar{\varepsilon}_j^2} \{\phi_j^r\}_q. \qquad [5.29]$$

Here, one-dimensional $\mathscr{C}^2$ cubic spline shape functions are chosen, though we may employ other more advanced interpolation schemes such as diffuse approximation [XIA 13a]. For a multidimensional strain domain, this reduced database model requires computing the coefficients of one-dimensional spline functions, which further reduces both computing time and operational complexity in the on-line phase.

5.4. Structural topology optimization

Substituting the local material optimization process by the NEXP approximated solution-dependent nonlinear material behavior, as shown in Figure 5.3, the original simultaneous two-scale design problem [5.1] can now be transformed back to a conventional monoscale nonlinear structural design problem

$$\max_{\rho \in \mathcal{A}_\rho} \min_{u \in \mathcal{U}} \left\{ \frac{1}{2} \int_{\Omega} \rho(x) C_{ijkh}(u) \frac{\partial u_i}{\partial x_j} \frac{\partial u_k}{\partial x_h} \mathrm{d}\Omega - l(u) \right\}, \qquad [5.30]$$

where elastic stiffness tensor $C_{ijkh}(u)$ is dependent on the displacement solution u. The explicit representation of this nonlinear behavior has been constructed in section 5.3 using the NEXP model.

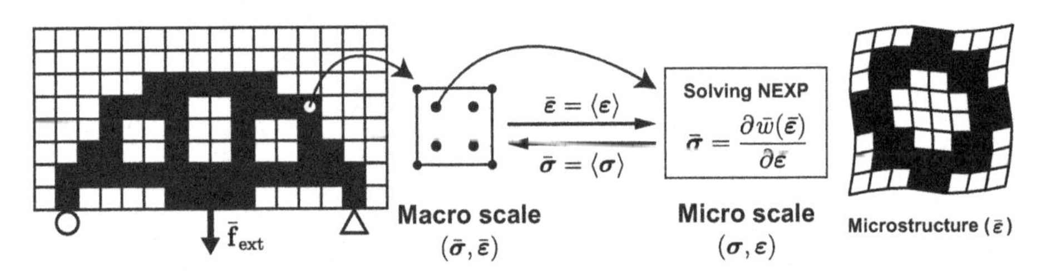

Figure 5.3. *Illustration of the monoscale structural design with the NEXP model*

In analogy to [4.9], the vector of topology design variables $\boldsymbol{\rho} = (\rho_1, \ldots, \rho_{N_s})^T$ is defined in an elementwise manner in association with the element internal force vector $\bar{\mathbf{f}}^i_{\text{int}}$

$$\bar{\mathbf{f}}^i_{\text{int}} = \rho_i \int_{\Omega_i} \bar{\mathbf{B}}^T \bar{\boldsymbol{\sigma}} \, \mathrm{d}\Omega_i, \quad i = 1, \ldots, N_s \qquad [5.31]$$

where the effective stress $\bar{\boldsymbol{\sigma}}$ is computed via the NEXP approximation and N_s is the number of discretized elements for the considered structure.

In analogy to [4.13], we can equivalently express [5.30] in a discretized form as the minimization of structural compliance [BEN 03]

$$\begin{aligned} \min_{\boldsymbol{\rho}} &: f_\mathrm{c}(\boldsymbol{\rho}, \bar{\mathbf{u}}) \\ \text{subject to} &: \bar{\mathbf{r}}(\bar{\mathbf{u}}, \boldsymbol{\rho}) = \mathbf{0} \\ &: V_\rho = \sum \rho_i v_i = V^\mathrm{s}_\mathrm{req} \\ &: \rho_i = 0 \text{ or } 1,\ i = 1, \ldots, N_\mathrm{s} \end{aligned} \qquad [5.32]$$

where $\bar{\mathbf{f}}_\mathrm{ext}$ is the external force vector, $V(\boldsymbol{\rho})$ and V^s_req are the total volume of solid elements and the required volume of solid elements, respectively, v_i is the volume of the ith element and $\bar{\mathbf{r}}(\bar{\mathbf{u}}, \boldsymbol{\rho})$ stands for the force residual at the macroscopic scale

$$\bar{\mathbf{r}}(\bar{\mathbf{u}}, \boldsymbol{\rho}) = \bar{\mathbf{f}}_\mathrm{ext} - \sum_{i=1}^{N} \rho_i \int_{\Omega_i} \mathbf{B}^T \bar{\boldsymbol{\sigma}} \mathrm{d}\Omega_i. \qquad [5.33]$$

where Ω_i denotes the region occupied by the ith element.

Similarly to [4.15], the sensitivity for the nonlinear design problem is evaluated according to the definitions in [5.32] and [5.33] as in [BEN 03]

$$\frac{\partial f_\mathrm{c}}{\partial \rho_i} = -\boldsymbol{\lambda}^T \int_{\Omega_i} \bar{\mathbf{B}}^T \bar{\boldsymbol{\sigma}} \mathrm{d}\Omega_i, \qquad [5.34]$$

where $\boldsymbol{\lambda}$ is the adjoint solution of

$$\bar{\mathbf{K}}_\mathrm{tan}(\bar{\mathbf{u}}_\mathrm{sol}) \boldsymbol{\lambda} = \bar{\mathbf{f}}_\mathrm{ext}. \qquad [5.35]$$

where $\bar{\mathbf{K}}_\mathrm{tan}$ is the tangent stiffness matrix at the convergence $\bar{\mathbf{u}}_\mathrm{sol}$ of the NR solution process. As has been shown in Figure 4.2, the tangent stiffness matrix for a certain displacement solution is the corresponding elastic stiffness matrix itself, i.e. $\bar{\mathbf{K}}_\mathrm{tan}(\bar{\mathbf{u}}_\mathrm{sol}) = \bar{\mathbf{K}}(\bar{\mathbf{u}}_\mathrm{sol})$ and therefore $\boldsymbol{\lambda} = \bar{\mathbf{u}}_\mathrm{sol}$. The evaluation of [5.34] can be further simplified to

$$\frac{\partial f_\mathrm{c}}{\partial \rho_i} = -\bar{\mathbf{u}}^T_\mathrm{sol} \int_{\Omega_i} \bar{\mathbf{B}}^T \bar{\boldsymbol{\sigma}} \mathrm{d}\Omega_i, \qquad [5.36]$$

which can be analytically approximated using the NEXP model

$$\frac{\partial f_c}{\partial \rho_i} \approx -\bar{\mathbf{u}}_{\text{sol}}^T \int_{\Omega_i} \bar{\mathbf{B}}^T \frac{\partial \tilde{w}(\bar{\varepsilon})}{\partial \bar{\varepsilon}} \mathrm{d}\Omega_i. \qquad [5.37]$$

5.5. General design algorithm

The general design algorithm for multiscale structural topology optimization consists of two phases. The off-line phase builds the approximate constitutive model $\tilde{w}(\bar{\varepsilon})$ for locally optimized materials following the NEXP strategy presented in section 5.3. The main steps involved in the off-line phase are summarized in algorithm 5.1. The on-line design phase follows the design framework presented in algorithm 4.1 except for the local material optimization process, which is now substituted by the NEXP approximation using the precomputed constitutive model $\tilde{w}(\bar{\varepsilon})$. The on-line design phase with NEXP approximation is summarized in algorithm 5.2.

Algorithm 5.1. NEXP database construction (off-line phase, section 5.3)

1: define the cellular material model;
2: define the discretized test domain $\Delta \in \mathbb{R}^D$;
3: solve local optimization [5.6] for each node of Δ;
4: store the strain energy density values into $\mathbb{W}$ as the database;
5: decompose $\mathbb{W}$ into a separated form in terms of $(\bar{\varepsilon}_1, \ldots, \bar{\varepsilon}_6)$;
6: build $\tilde{w}(\bar{\varepsilon})$ as a sum of products of the 1D interpolation functions.

Algorithm 5.2 eventually gives the optimal structural topology at the structural scale, where the corresponding local optimal material topologies are not yet obtained. In order to retrieve local material topologies, one just needs to perform local material topology optimizations [XIA 14a, XIA 15a] using the converged solution at the final converged structural topology. This two-scale design strategy requires significantly less computational efforts than the simultaneous design strategy discussed in Chapter 4.

5.6. Numerical examples

In this section, several numerical examples are considered. In section 5.6.1, the NEXP model is constructed for a considered local material design problem and preliminary tests on its performance are given. In section 5.6.2, the design of the two-scale bridge-type structure (section 5.3 in [XIA 14a]) is revisited for a further validation of the constructed NEXP model. Two two-scale structures with fine discretization are considered for design following algorithm 5.2 in sections 5.6.3 and 5.6.4 using the constructed NEXP model in section 5.6.1.

Algorithm 5.2. Topology optimization (on-line phase, section 5.4)

1: initialize ρ_0 and $\mathbf{K}_0$;
2: **while** $\|\rho_{i+1} - \rho_i\| > \delta_{\text{opt}}$ **do**
3:     **while** $\|\bar{\mathbf{f}}_{\text{ext}} - \bar{\mathbf{f}}_{\text{int}}\| > \delta_{\text{f}}$ **do**
4:         **loop** over all macro Gauss points
5:             compute the effective strain $\bar{\varepsilon}$;
6:             compute the effective stress $\bar{\boldsymbol{\sigma}} \approx \partial\tilde{w}/\partial\bar{\varepsilon}$;
7:         **end loop**
8:         NR update: $\bar{\mathbf{K}}_0\Delta\bar{\mathbf{u}} = \bar{\mathbf{f}}_{\text{ext}} - \sum \rho_i \int_{\Omega_i} \bar{\mathbf{B}}^T\bar{\boldsymbol{\sigma}}\mathrm{d}\Omega_i$;
9:     **end while**
10:     compute f_{c} and sensitivities $\partial f_{\text{c}}/\partial\rho$;
11:     update ρ using BESO scheme;
12: **end while**
13: **return** ρ.

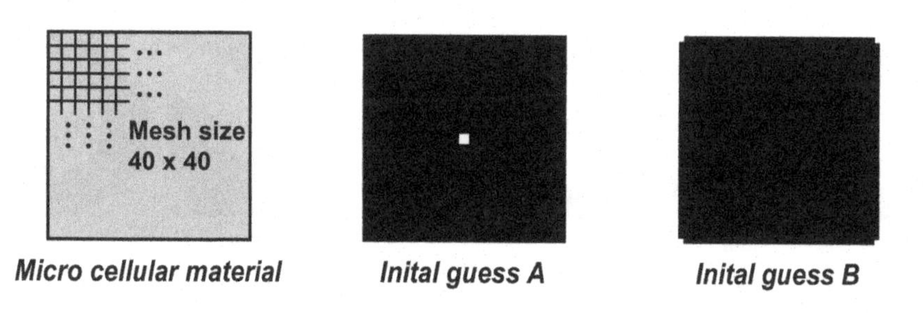

Figure 5.4. *Illustration of the cellular material model and two initial material guesses*

5.6.1. *Preliminary numerical tests*

First, we consider a discretized cellular material model at the microscopic scale defined, as shown in Figure 5.4. The square RVE of cellular material is discretized into 40×40 4-node plane stress bilinear elements and the same number of density design variables is correspondingly defined. Young's modulus and Poisson's ratio of the solid material are set to 1 and 0.3, respectively. The volume fraction of the solid material within the microstructure is set to 60%, i.e. a microporosity of 40% is assumed. By the BESO method, redundant or inefficient material is gradually removed from the structure until the prescribed volume constraint is reached. The evolution rate in the BESO is set to $c_{\text{er}} = 0.02$, which determines the percentage of removed material at each design iteration. In order to obtain the so-called one-length scale microstructure [BEN 03], i.e. avoid too detailed microstructures inside the cell, sensitivities are filtered within a local zone controlled by a filter radius $r_{\text{min}} = 6$. Following [HUA 11, HUA 12], two initial material guesses are considered for design. Initial guess A assigns four soft elements at the center of the design domain to trigger topological changes. Initial guess B assigns four soft elements at the four corners. The material microstructures optimized from initial guess A and B for several representative loading cases are shown in Figures 5.5 and 5.6. As argued in [BEN 03], different initial guesses may lead to different microstructures that may possess similar material properties due to the non-uniqueness of the solution. Sometimes the microstructural topologies obtained from different initial guesses are in fact the shifted versions of the same topology. In the following tests, all material microstructures are designed from initial guess A.

Figure 5.5. *Several representative material designs from initial guess A for loading cases: $\bar{\varepsilon} = (1.0, 0, 0)^T$, $(1.0, 0.8, 0)^T$, $(1.0, 1.0, 0)^T$, $(1.0, 0.7, 0.8)^T$, $(0, 0, 1.0)^T$*

The NEXP model is built over the strain domain $\Delta = \Delta_1 \times \Delta_2 \times \Delta_6 = [-1, 1] \times [-1, 1] \times [-1, 1]$ in the 2D case. After comparing different test results and with the consideration of implementation efficiency, we choose to discretize each dimension of the strain space into $p = 21$ uniformly distributed points, which means in total 21^3 local material optimizations are performed. Detailed comparison of the NEXP models using different discretization choices can be found in [YVO 09, YVO 13]. Following the algorithm 5.1, an explicitly represented strain energy density and effective strain relationship is constructed. With a relative reconstruction error in [5.22] chosen as 0.01, we obtain $R = 9$ truncated modes in each dimension for the reduced approximation $\tilde{w}(\bar{\varepsilon}_1, \bar{\varepsilon}_2, \bar{\varepsilon}_6)$. It is worth noting that when the cellular material design problem is symmetrically defined for both its geometry and boundary conditions, the decomposed modes should possess similar symmetric features. The first five of the nine normalized modes are selected and illustrated in Figure 5.7 with expected symmetries. Such symmetries help reduce the number of precomputations during the off-line phase, in particular for 3D problems. In this 2D case, the symmetry of $\bar{\varepsilon}_6$ reduces the number of precomputations by half. The bisymmetry of $\bar{\varepsilon}_1$ and $\bar{\varepsilon}_2$ further reduces the number of precomputations to one-fourth. Therefore, in total only one-eighth precomputations are required in 2D case for the construction of the database.

Figure 5.6. *Several representative material designs from initial guess B for loading cases:* $\bar{\varepsilon} = (1.0, 0, 0)^T$, $(1.0, 0.8, 0)^T$, $(1.0, 1.0, 0)^T$, $(1.0, 0.7, 0.8)^T$, $(0, 0, 1.0)^T$

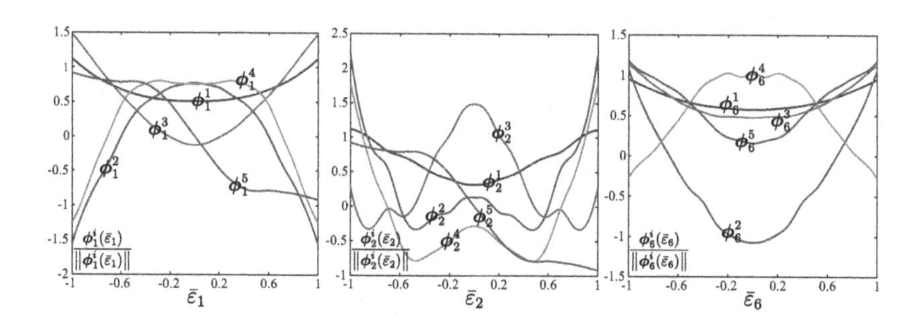

Figure 5.7. *The first five normalized modes for the three strain components. For a color version of the figure, see www.iste.co.uk/xia/topology.zip*

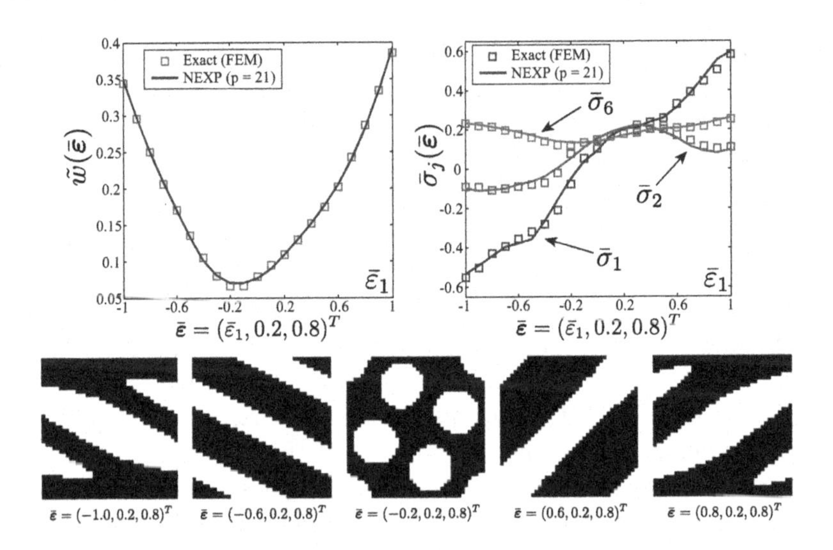

Figure 5.8. *Comparison case 1 of the exact and approximate values evaluated using FEM and NEXP. For a color version of the figure, see www.iste.co.uk/xia/topology.zip*

To validate the performance, we compare the values evaluated using the NEXP model and the exact computations. A first comparison is given in Figure 5.8, where $\bar{\varepsilon}_2 = 0.2$, $\bar{\varepsilon}_6 = 0.8$ are fixed and $\bar{\varepsilon}_1$ varies from -1 to 1. Several observations can be found in Figure 5.8. First, the approximate values given by the NEXP model are in very good agreement with the exact solutions. Second, the strain energy density is a convex function over the effective strain space. Third, the selected optimal material microstructures show the tendency of topological variation along $\bar{\varepsilon}_1$, which is the reason for the nonlinearity of [5.30]. Another validation is given in Figure 5.9, where $\bar{\varepsilon}_1 = -0.1$, $\bar{\varepsilon}_6 = 0.5$ are fixed and $\bar{\varepsilon}_2$ varies from -1 to 1. Similar observations can be found in Figure 5.9 as shown in Figure 5.8. Note that, since the tangent stiffness matrix is not required in the present work (see algorithm 5.2), we did not validate the second-order derivatives of the NEXP model.

5.6.2. *Design of a two-scale bridge-type structure*

To further validate the NEXP model constructed in section 5.6.1, we revisit the two-scale bridge-type structure design problem in section 4.4.3 as shown in Figure 5.10. The macroscopic structure is discretized into 40×20 4-node bilinear elements and the same number of topology design variables are correspondingly defined. Each element possesses four Gauss integration points, which means in total $N_{\text{gp}} = 4 \times 40 \times 20 = 3,200$ cellular material models with 40% porosity are considered at the microscopic scale. The same cellular material model is considered as given in section 5.6.1.

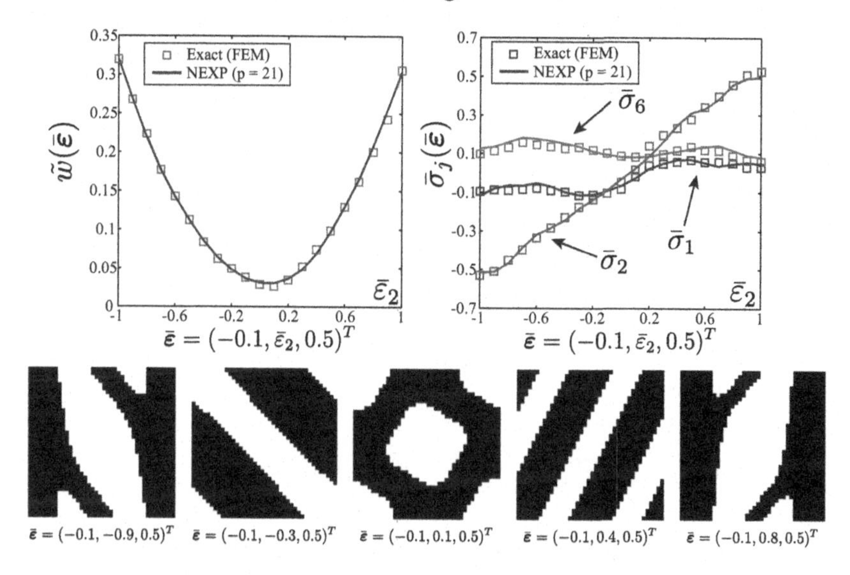

Figure 5.9. *Comparison case 2 of the exact and approximate values evaluated using FEM and NEXP. For a color version of the figure, see www.iste.co.uk/xia/topology.zip*

Before comparing optimal structural topologies, we first compared the equilibrium convergence for pure structural analysis using the simultaneous design strategy in Chapter 4 and the reduced strategy currently proposed. With the displacement convergence criterion set as $\delta = \|\bar{\mathbf{u}}^{(k+1)} - \bar{\mathbf{u}}^{(k)}\|_2 / \|\bar{\mathbf{u}}^{(k)}\|_2 \leq 10^{-2}$, it took seven substeps to reach the macroscopic structural equilibrium using the unreduced strategy at a cost of 10 min computing for each substep on a personal computer. In contrast, the reduced approach requires only 30 sec for each substep computation and gives a similar convergence history as shown in Figure 5.11. Following the general design algorithm presented in section 5.5, the retrieved optimal cellular material topologies at the convergence obtained by the reduced strategy are given in Figure 5.12(a). It is important to emphasize that the optimized cellular material topologies only represent the optimal solutions at the microscopic scale for the associated material point satisfying the assumptions of scale separation and periodicity. The optimized cellular materials in neighboring Gauss points are not necessarily contiguous with each other. For the purpose of comparison, the optimal cellular material topologies at the convergence obtained by the unreduced strategy are shown in Figure 5.12(b). Though the unreduced strategy gives more complex material microstructures than the reduced strategy in certain local regions, two topologies in Figure 5.12 in general have very similar structural tendencies and their structural compliance values are also very close.

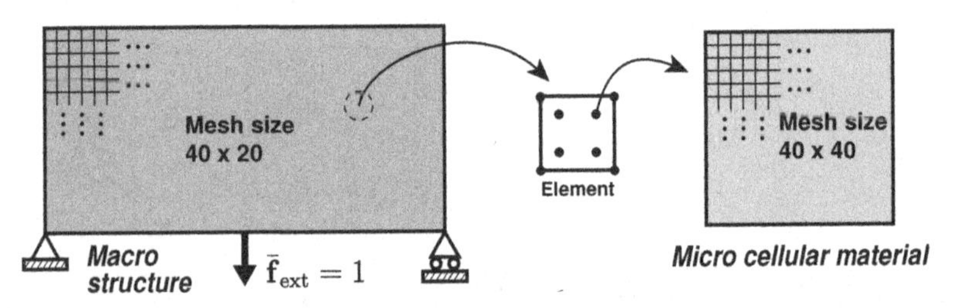

Figure 5.10. *Illustration of the two-scale bridge-type structure (Figure 4.7)*

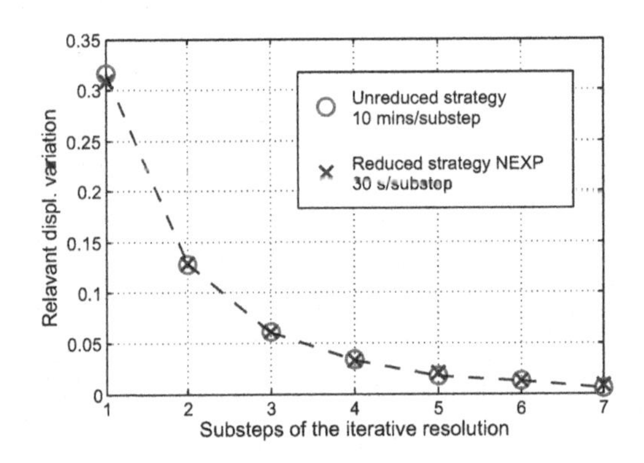

Figure 5.11. *Comparison of the convergence histories of unreduced and reduced strategies. For a color version of the figure, see www.iste.co.uk/xia/topology.zip*

In the next step, topology optimization is performed at the macroscopic structural scale using the NEXP model. Volume constraint at the macroscopic scale is also set to 60%. The other related parameters in the BESO method are the evolution rate $c_{\text{er}} = 0.02$ and filter radius $r_{\text{min}} = 3$. From the BESO method, it takes 39 design iterations to reach the final design as shown in Figure 5.13(a), where each design iteration requires seven substeps to reach the equilibrium. For the purpose of comparison, Figure 5.13(b) gives the two-scale designed topology obtained using the unreduced strategy (Chapter 4). As expected, an obvious topological difference between the two designs can be observed in Figure 5.13. Note that Figures 5.8 and 5.9 are just two of the selected validation cases for the NEXP approximation. The approximation accuracy cannot be guaranteed for all possible cases as it is shown in Figures 5.8 and 5.9. Moreover, even slight approximation errors result in a variation in the structural performance, which leads to a different topological evolution.

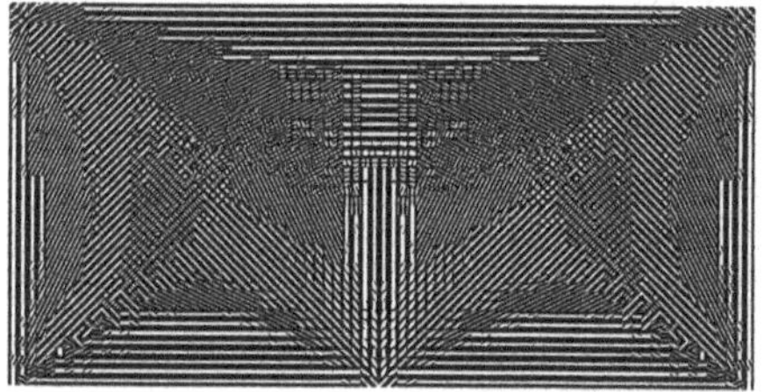

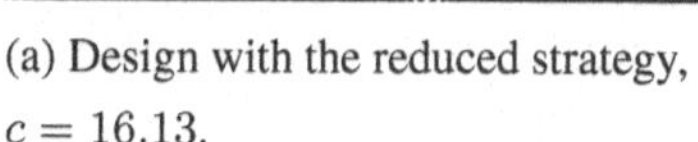

(a) Design with the reduced strategy, $c = 16.13$.

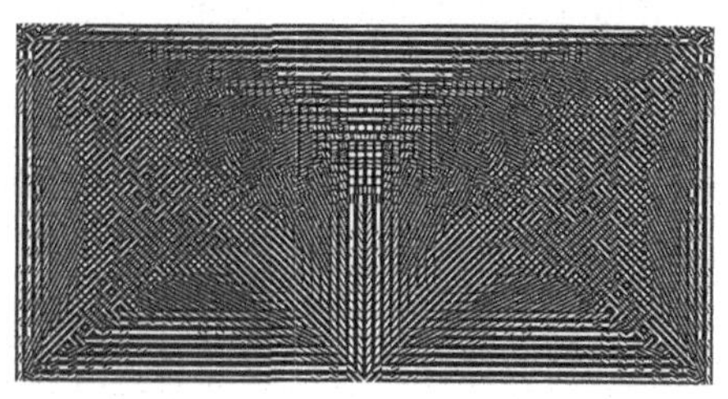

(b) Design with the unreduced strategy, $c = 16.21$.

Figure 5.12. *Comparison of the material designs using reduced and unreduced strategies*

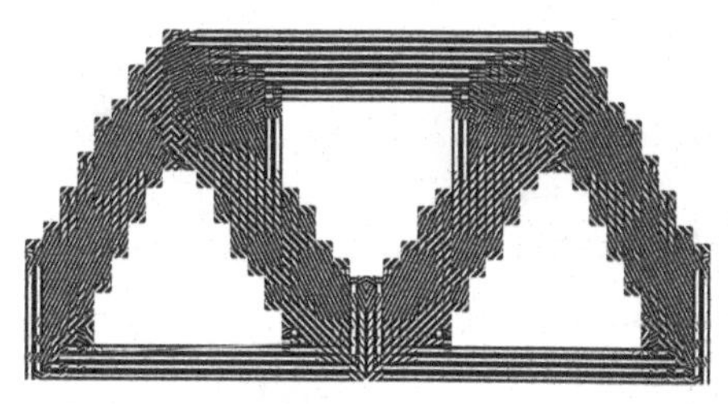

(a) Design with the reduced strategy, $c = 20.09$.

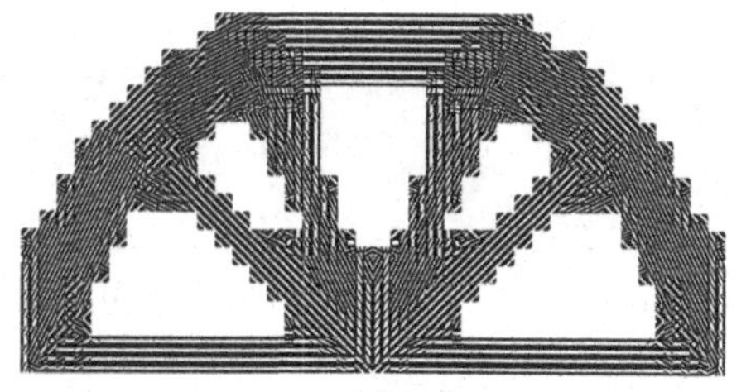

(b) Design with the unreduced strategy, $c = 20.23$.

Figure 5.13. *Comparison of the two-scale topologies using reduced and unreduced strategies*

A detailed comparison of the two topology optimization procedures is shown in Figure 5.14. The two topology evolutions at the structural scale follows similar topological variations at the beginning iterations. The first obvious difference appears at iteration 13, where the unreduced strategy initiates two holes inside the two inner branches (Figure 5.14(d)) while the reduced strategy does not initiate any such process (Figure 5.14(a)). Such differences are reasonable due to the NEXP approximation error. The two holes initiated by the unreduced strategy grow progressively (Figures5.14(e) and (f)) along with the subsequent optimization process and eventually converge to Figure 5.13(b). On the other hand, the initiated holes in the latter iteration (Figures 5.14(b) and (c)) by the reduced strategy eventually disappear and the structure converges to another local optima (Figure 5.13(a)). As can be observed from Figures 5.13 and 5.14, the reduced strategy unexpectedly produces design solutions with even lower compliance values than those obtained by the unreduced strategy. To the best of our knowledge, the reason may be twofold. First, this phenomenon may be due to the numerical issues in topology optimization procedure, where Figure 5.13(a) is the converged design result at iteration 39, whereas Figure 5.13(b) is the converged design result at iteration 31. It is natural to have an improved objective value with additional design iterations. Second, another reason for this phenomenon is due to the NEXP approximation error, which indicates that the NEXP approximated material is slightly stiffer than the real material behavior. The above results also imply that though the NEXP model may not perfectly replace the nonlinear behavior of local material optimization, it can still produce local optimal designs at an extremely reduced computational cost.

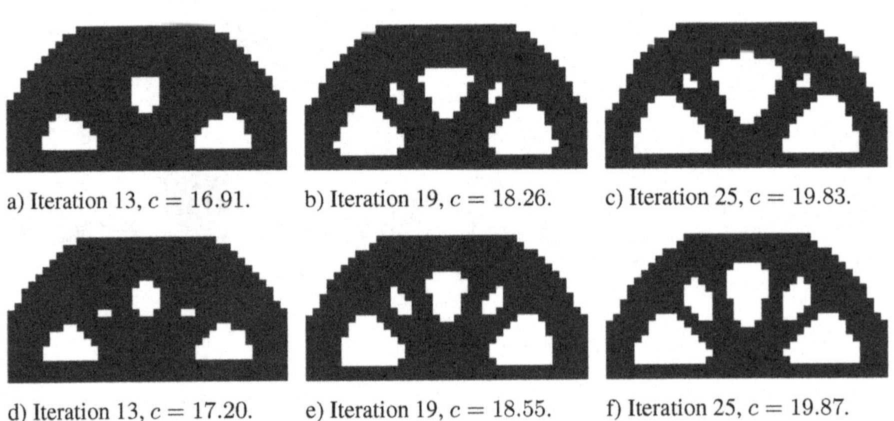

a) Iteration 13, $c = 16.91$. b) Iteration 19, $c = 18.26$. c) Iteration 25, $c = 19.83$.

d) Iteration 13, $c = 17.20$. e) Iteration 19, $c = 18.55$. f) Iteration 25, $c = 19.87$.

Figure 5.14. *Intermediate iterations of reduced (a, b and c) and unreduced (d, e and f) strategies*

5.6.3. *Design of a finely discretized two-scale bridge-type structure*

With the same NEXP model to approximate the nonlinear material behavior of local material optimizations, we are now capable of designing a much larger scale or finely discretized two-scale bridge-type structure of the previous example as shown in Figure 5.10. The macroscopic structure is discretized into 120×60 4-node bilinear elements and the same number of topology design variables are correspondingly defined. The same cellular material model with 60% material usage, i.e. 40% porosity as considered in the first example, is attributed to each macroscopic integration point.

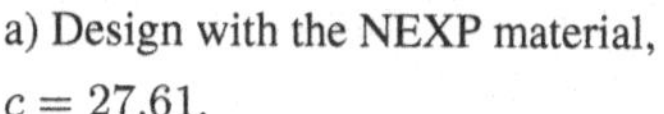

a) Design with the NEXP material, $c = 27.61$.

b) Design with the HS upper bound, $c = 37.85$.

Figure 5.15. *Comparison of topology designs using the NEXP material and the SIMP material*

Topology optimization is performed at the macroscopic structural scale using the NEXP material model. Volume constraint at the macroscopic structure is also set to 60%. The other related parameters in the BESO method are the evolution rate $c_{\text{er}} = 0.02$ and filter radius $r_{\min} = 5$. From the BESO method, it takes 37 design iterations to reach the final design as shown in Figure 5.15(a), where each design iteration requires six substeps to reach the equilibrium with the same criterion discussed in section 5.6.2. The considered structure is too finely discretized to be optimized using the simultaneous design strategy [XIA 14a] (in total $120 \times 60 \times 4$ local problems); for this reason, we only give the linear design result using isotropic porous material with 40% porosity, as shown in Figure 5.15(b), for comparison. Following [BEN 99], the effective Young modulus and Poisson ratio for the isotropic porous material with 40% porosity obtained by inverse homogenization [SIG 00] corresponding to the Hashin–Shtrikman (HS) upper bound equal to 0.34 and 0.3, respectively. The NEXP model approximates the behavior of an anisotropic material with 40% material porosity, which has a maximized strain energy value for the given strain status. As expected, structure, as shown in Figure 5.15(a), optimized using the NEXP model has a much lower compliance than the structure, as shown in Figure 5.15(b), optimized using the assumed porous material corresponding to the HS upper bound. At the same time, an obvious difference can be observed from the two topologies, as shown in Figure 5.15, which again implies the necessity of two-scale material and structure design.

The optimized structure topology together with its retrieved optimal cellular material topologies is shown in Figure 5.16. Due to the fine discretization at the structural scale, the detailed optimal microstructures can barely be observed from the retrieved two-scale topology. For a better visualization of microscopic material topology layout, we have selected and zoomed in on three local zones as shown in Figure 5.16. As can be observed, uniaxial materials may be sufficient at the main branches of the structure; while in order to have a better structural performance, anisotropic materials have to be used at the joints of the main branches due to the more complex loading status. It takes 10 h for all 37 iterations on the personal computer for solving such a large-scale problem with the NEXP model. Retrieving microscopic material topologies at the final converged solution itself requires an additional 2 h computing time. Assuming six substeps required on average for each structural design iteration and 2 h computing time for each substep of each design iteration, the simultaneous design strategy (Chapter 4) would require in total more than 400 h ($37 \times 6 \times 2$) computing time, which is obviously not affordable. It is important to notice that the local material optimization problems are independent from each other and we can further reduce the computing time by combining the reduced strategy and parallel computing.

5.6.4. *Design of a two-scale half MBB beam with fine discretization*

Another benchmark problem, the so-called MBB beam [BEN 03], is further considered in this example. Similar tests have also been investigated in [ROD 02, XIA 14a] for the simultaneous material and structure design, and recently in [GAO 12] for composite laminate orientation design. Due to the symmetry of the problem, only a half MBB beam is considered, as shown in Figure 5.17. The macroscopic structure is discretized into 100×40 4-node bilinear elements. The same cellular material model with 40% porosity as discussed in section 5.6.1 is attributed to each macroscopic integration point, in total $4 \times 100 \times 40 = 16{,}000$ cellular material models are considered at the microscopic scale.

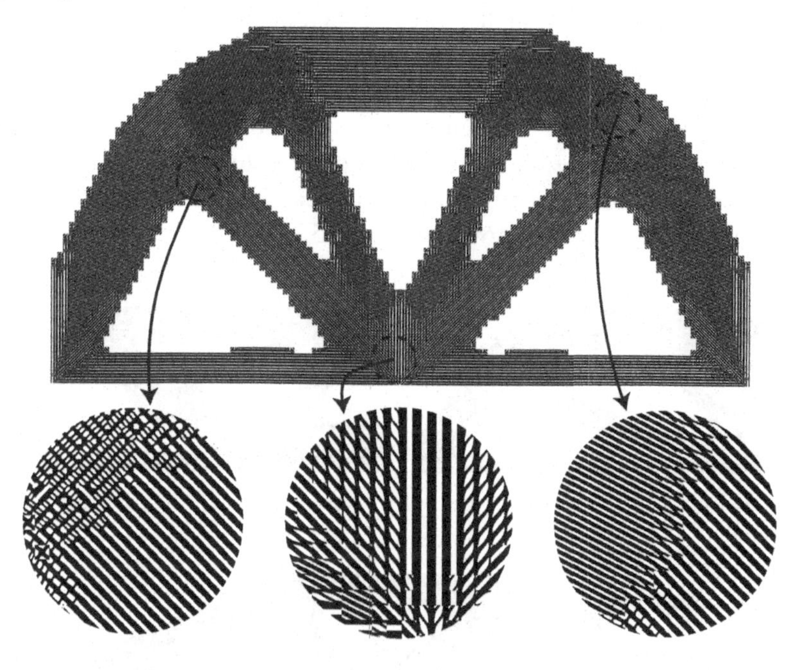

Figure 5.16. *Design of bridge-type structure with retrieved local optimal material topologies*

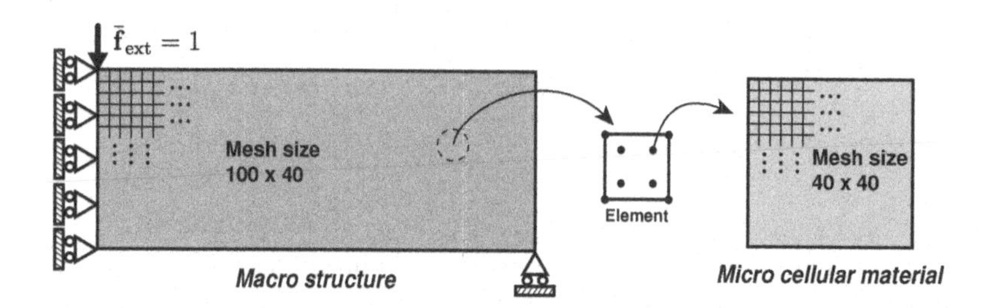

Figure 5.17. *Illustration of a finely discretized two-scale half MBB beam*

Similar to the previous example, topology optimization is performed at the macroscopic structural scale using the NEXP model. Volume constraint at the macroscopic structure is also set to 60%. The other related parameters in the BESO algorithm are the evolution rate $c_{\text{er}} = 0.02$ and filter radius $r_{\min} = 5$. Figure 5.18(a) gives the converged design topology at the 35 iteration. It requires six substeps on average for each design iteration to reach the equilibrium with the criterion discussed in section 5.6.2. Comparison designs are shown in Figures 5.18(b) and (c), where Figure 5.18(b) is constituted of the isotropic material with 40% porosity corresponding to the upper HS bound [BEN 99] and the structural topology shown in Figure 5.18(c) is constituted of a prescribed holed microstructure with 40% porosity assumed at the microscopic scale. Obvious differences can be observed from the three topologies and the structure shown in Figure 5.18(a) is the stiffest among the three.

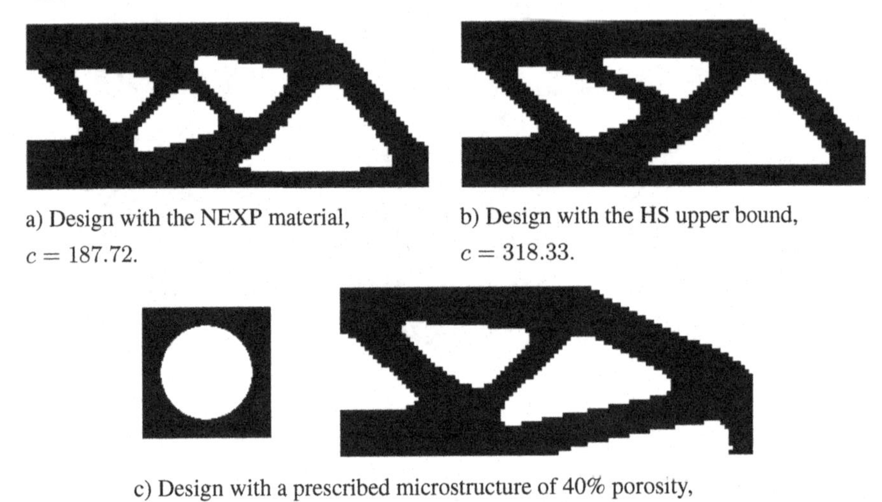

a) Design with the NEXP material, $c = 187.72$.

b) Design with the HS upper bound, $c = 318.33$.

c) Design with a prescribed microstructure of 40% porosity, $c = 359.25$.

Figure 5.18. *Comparison of topology designs using the different material models*

Figure 5.19 gives the optimized structure topology together with its retrieved optimal cellular material topologies. Three local zones are selected and zoomed in on for a better visualization of microscopic material topologies. This test takes around 6 h for all 35 iterations on the personal computer. Retrieving microscopic material topologies at the final converged solution requires one additional hour of computing. Assuming six substeps are required on average for each structural design iteration and 1 h of computing for each substep of each design iteration, the simultaneous design strategy [XIA 14a] would require in total more than 200 h of computing time for solving this problem on the personal computer. In contrast, using the current strategy with an approximate NEXP material model, it requires only 7 h computing to reach the final two-scale topology design, which can still be further reduced with parallel computing.

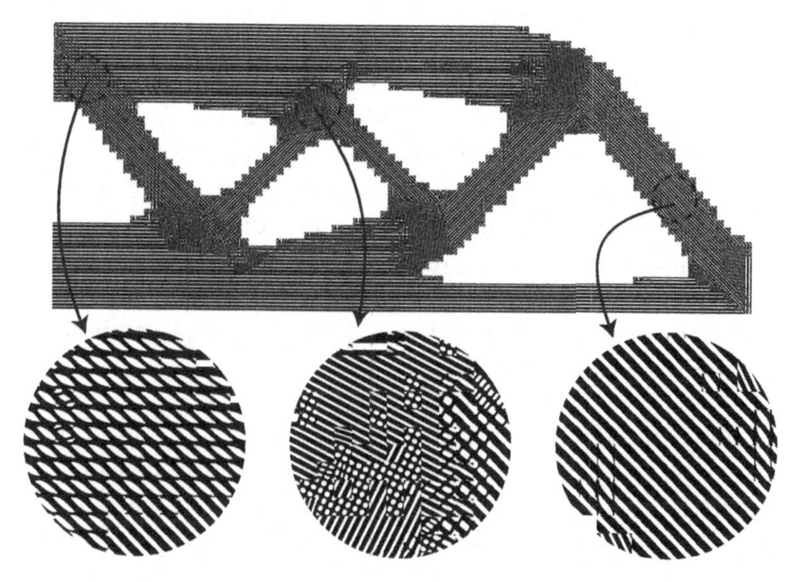

Figure 5.19. *Design of half MBB beam with retrieved local optimal material topologies*

5.7. Concluding remarks

In this chapter, we have proposed using the NEXP model to approximate the nonlinear behavior of locally optimized or adapted materials within the developed simultaneous design framework. It has been shown that this explicit NEXP approximation can serve the simultaneous design purpose well, providing ultraresolution structures at a significantly reduced computational cost.

This work extends the NEXP model to a new regime of nonlinearity. At the same time, this work inherits limitations of the NEXP model, such as the choices of test space discretization, interpolation scheme and tensor decomposition strategy. The computational cost in the off-line and on-line phases linearly depends on the number of points in the database and the number of integration points in the considered structure, respectively. It is important to note that the local material optimization problems are independent, demonstrating that the tasks are largely parallel. By simultaneously using parallel computing and the NEXP model (as we did in Chapter 3), further reduction can be straightforwardly achieved.

The proposed simultaneous design framework is in general independent of the type of design variables and the optimization model. Other types of design variables such as geometrical or even manufacturing process parameters can be considered instead of topology design variables for both scales. More realistic material models may further be considered at the microscopic scale with the consideration of more complex material and geometrical nonlinearities. It would be highly promising that the NEXP model has a good performance in approximating the nonlinearity of local optimization or adaptation of more realistic materials.

In addition, the decoupled approach using the NEXP model also suits other macroscopically defined objectives, such as displacements or stresses. As opposed to the macroscopically defined objectives, when the design objective is defined locally at the microscopic scale, the decoupled approach encounters difficulties and more specifically developed ROMs are required.

Conclusion and Perspectives

In this book, we have developed first in Chapter 1 a multiscale design framework consisted of two components: *topology optimization* and *multiscale modeling*. In contrast to the conventional nonlinear design of homogeneous structures, this design framework provides an automatic design tool for nonlinear highly heterogeneous structures of which the underlying material model is governed directly by realistic microstructural geometry and microscopic constitutive laws.

With regard to the computational and data storage requirements due to multiple realizations of multiscale computing, we have introduced to the design framework a third ingredient: *reduced-order modeling*. We have developed in Chapter 2 an adaptive surrogate model for the solutions at the microscopic scale, which has shown promising performance when applied to the design framework for nonlinear elastic cases. As for more severe material nonlinearity, we have employed in Chapter 3 the pRBMOR model with GPU acceleration, which allows us to realize the design of multiscale elastoviscoplastic structures in realistic computing times and with affordable memory requirements. Note that without *reduced-order modeling*, the computational investment required for such designs is beyond current (and likely future) capabilities. Hence, the proposed multiscale design framework allows us to tackle problems that were until recently unsolveable.

In pursuing higher performance structures, in Chapter 4 we extended the multiscale design framework by introducing additional design variables at the microscopic scale to perform simultaneous design of structure and material microstructures. In particular, we treat the material optimization process integrally as a generalized nonlinear constitutive behavior and propose an initial stiffness NR solution scheme to resolve this specific nonlinear equilibrium problem. The proposed model allows us to obtain optimal structures with spatially varying properties realized by the simultaneous design of microstructures, which greatly favors the 3D printing setting that a single material can usually be used for fabrication. We have improved the design efficiency in Chapter 5 by a straightforward application of the NEXP model to approximate the generalized nonlinear constitutive behavior. It has been shown that the explicit NEXP approximation can serve the simultaneous design purpose of providing ultraresolution structures at a significantly reduced computational cost.

To the best of our knowledge, design of multiscale structures is a relatively new field in which there has been very limited research, especially for nonlinear cases. Many potential developments for the proposed multiscale design framework can be carried out with respect to any of the three components: *topology optimization*, *multiscale modeling* and *reduced-order modeling*. In the following, we give our perspectives on potential extensions based on the proposed multiscale design framework:

1) As can be observed from Figure 1.10, the presence of microstructures results in high concentrations with stress values much higher than the effective values evaluated at the macroscopic scale. These stress concentrations may result in the initial material failure or crack at the microscopic scale. Therefore, there is a necessity to limit or constrain the maximum stress values at the microscopic scale rather than at the macroscopic scale when it comes to the design of heterogeneous structures such as laminated composite, concrete and alloy structures.

2) In Chapters 1–3, we have assumed a fixed RVE at the microscopic scale. By the developed multiscale design framework, we can consider as well a set of RVEs with variant microstructures and material compositions. As for model reduction, it would require an individual ROM for each considered RVE. The ultimate goal would be simultaneous design of structure and the underlying RVEs in terms of geometrical or even manufacturing process parameters. Nevertheless, developing an individual ROM for each of the feasible RVEs is obviously impractical due to the enormous number of potential RVEs. In such cases, one potential strategy is to develop specific parameterized ROMs to serve the design requirements.

3) As was discussed in section 4.4.3 and can be seen from Figure 4.11(a), the connectivity among the optimized material microstructures is not guaranteed due to the assumption of a clear separation of scales. Such designs with unconnected microstructures are not manufacturable from a practical point of view. This could be alleviated by imposing an artificial length ratio and connect the optimized microstructures at the neighboring Gauss points by certain transitional microstructures; however, the expected mechanical performance is consequently no longer guaranteed. Alternatively, we may apply more sophisticated so-called marco-meso models by implementing higher order homogenization schemes, where the length ratio is explicitly defined.

Appendix

Design of Extreme Materials in Matlab

A.1. Introduction

Topology optimization [BEN 88] was first employed for the material design by Sigmund [SIG 94] via an inverse homogenization approach. This was followed by a series of systematic works [SIG 97, SIG 00, GIB 00]. The subject has been later successively addressed by the density-based approach [NEV 00, GUE 07, ZHA 07], the level set method [CHA 08, WAN 14c], the topological derivative [AMS 10] and the ESO-type method [HUA 11]. Figures A.1–A.3 show some representative extremal microstructures designed by topology optimization. Functionally graded material and structure designs have been given by Paulino *et al.* [PAU 09] and Almeida *et al.* [ALM 10]. Another closely related area of research is concurrent material and structural design [ROD 02, ZHA 06, XIA 14a, XIA 15b].

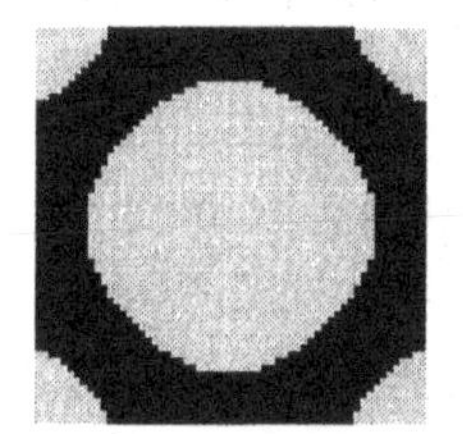  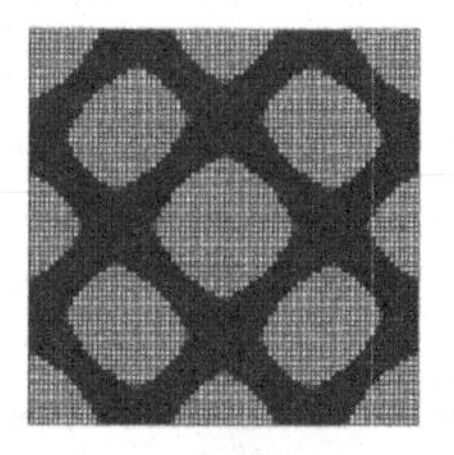 

Figure A.1. *Microstructures with maximized bulk moduli: the first two from [SIG 00], the third from [ZHA 07] and the last from [AMS 10] (from left to right)*

Figure A.2. *Microstructures with maximized shear moduli: the first two from [NEV 00], the third from [HUA 11] and the last from [AMS 10] (from left to right)*

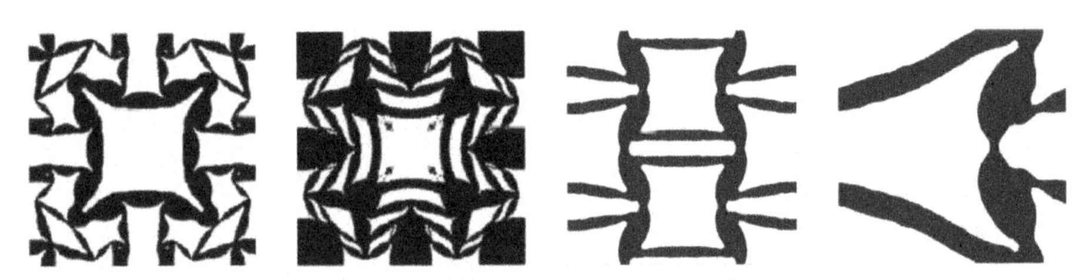

Figure A.3. *Microstructures with minimized negative Poisson's ratios: the first two from [AMS 10] and the last two from [WAN 14c] (from left to right)*

After the 99-line Matlab code in the seminal article by Sigmund [SIG 01], a series of educational papers with compact Matlab implementations have significantly contributed to the popularity and the development of topology optimization. These include a coupled level set method using the FEMLAB package by Liu *et al.* [LIU 05], the ESO method by Huang and Xie [HUA 10a], the discrete level set method by Challis [CHA 10], the 199-line code for Pareto-optimal tracing with the aid of topological derivatives by Suresh [SUR 10], the 88-line Matlab code by Andreassen *et al.* [AND 11], the Matlab code for the generation of polygonal meshes (PolyMesher) and the topology optimization framework (PolyTop) that are based on [TAL 12b, TAL 12a] and a parallel computing implementation [MAH 06].

We have also benefited from these educational papers. For instance, the multicomponent structural system designs [XIA 12, XIA 13b] are given within the framework of the 99-line code [SIG 01]. The reduced multiscale topology optimization [XIA 14b] uses the discrete level set method [CHA 10]. Moreover, our recent work on concurrent material and structural design [XIA 14a, XIA 15b] builds on top of the 88-line code framework [AND 11] along with the ESO optimizer [HUA 10a].

The present work extends the 88-line code to the optimal design of materials with extreme properties. We follow the design strategy proposed by Sigmund [SIG 94], where the homogenized material constitutive parameters are evaluated in terms of element mutual energies. For effective material property prediction, rather than using the conventional asymptotic expansion [GUE 90], we adopt an equivalent energy-based homogenization approach that employs average stress and strain theorems [HAS 83]. It will be shown in section A.6 that the applied design algorithm with the Matlab implementation (see section A.8) can generate extremal microstructures with similar topology configurations as shown in Figures A.1–A.3.

The remainder of the Appendix is organized as follows: in section A.2, homogenization theory is briefly reviewed. Section A.3 presents the implementation of periodic boundary conditions. Section A.4 gives the optimization model. Section A.8 explains the Matlab implementation. Section A.6 gives several numerical examples using the proposed code. Conclusions are drawn in section A.7 and the Matlab implementation is given in section A.8.

A.2. Homogenization

Within the scope of linear elasticity, the equivalent constitutive behavior of periodically patterned microstructures (Figure A.4) can be evaluated using the homogenization method [GUE 90]. Consider a single cell Y in $\mathbb{R}^3$

$$Y =]0, y_1^0[\times]0, y_2^0[\times]0, y_3^0[\,, \qquad \text{[A.1]}$$

where y_1^0, y_2^0 and y_3^0 are the dimensions of the base cell in the three directions.

Following asymptotic homogenization, the macroscale displacement field $u^\epsilon(x)$ depending on the aspect ratio ϵ between the macro- and microscales is expanded as

$$u^\epsilon(x) = u_0(x, y) + \epsilon u_1(x, y) + \epsilon^2 u_2(x, y) \ldots, y = x/\epsilon, \qquad \text{[A.2]}$$

where the involved functions are dependent on the global macroscopic variable x and the local microscopic variable y. The dependence on $y = x/\epsilon$ implies that a quantity varies within a very small neighborhood of a macroscopic point x, which may be viewed as "stretching" the microscale so it becomes comparable to the macroscale. When $\epsilon \ll 1$, the dependence on y can be considered periodic for a fixed macroscopic point x.

When only the first-order terms of the asymptotic expansion in equation [A.2] are considered, the homogenized stiffness tensor E^H_{ijkl} is given by averaging the integral over the base cell Y as

$$E^H_{ijkl} = \frac{1}{|Y|}\int_Y E_{ijpq}(\varepsilon^{0(kl)}_{pq} - \varepsilon^{*(kl)}_{pq})\mathrm{d}Y, \qquad \text{[A.3]}$$

where the Einstein index summation notation is used and $\varepsilon^{*(kl)}_{pq}$ is the Y-periodic solution of

$$\int_Y E_{ijpq}\varepsilon^{*(kl)}_{pq}\frac{\partial v_i}{\partial y_j}\mathrm{d}Y = \int_Y E_{ijpq}\varepsilon^{0(kl)}_{pq}\frac{\partial v_i}{\partial y_j}\mathrm{d}Y, \qquad \text{[A.4]}$$

where v is Y-periodic admissible displacement field and $\varepsilon^{0(kl)}_{pq}$ corresponds to the three (2D) or six (3D) linearly independent unit test strain fields.

With the intention of presenting a compact Matlab code, the energy-based approach that employs average stress and strain theorems is adopted in this work instead of the asymptotic approach. The energy-based approach imposes the unit test strains directly on the boundaries of the base cell, inducing $\varepsilon^{A(kl)}_{pq}$ that corresponds to the superimposed strain fields $(\varepsilon^{0(kl)}_{pq} - \varepsilon^{*(kl)}_{pq})$ in equation [A.3]. According to [HAS 83], these are two equivalent approaches for the prediction of material effective properties. A detailed implementation of periodic boundary conditions is given in section A.3.

With the intension to favor effective existing algorithms used in topology optimization, equation [A.3] is rewritten in an equivalent form in terms of element mutual energies [SIG 94]

$$E^H_{ijkl} = \frac{1}{|Y|}\int_Y E_{pqrs}\varepsilon^{A(ij)}_{pq}\varepsilon^{A(kl)}_{rs}\mathrm{d}Y. \qquad \text{[A.5]}$$

In finite element analysis, the base cell is discretized into N finite elements and [A.5] is approximated by

$$E^H_{ijkl} = \frac{1}{|Y|}\sum_{e=1}^{N}(\mathbf{u}_e^{A(ij)})^T\mathbf{k}_e\mathbf{u}_e^{A(kl)}, \quad \text{[A.6]}$$

where $\mathbf{u}_e^{A(kl)}$ are the element displacement solutions corresponding to the unit test strain fields $\boldsymbol{\varepsilon}^{0(kl)}$, and $\mathbf{k}_e$ is the element stiffness matrix. In 2D cases, we note that $11 \rightarrow 1$, $22 \rightarrow 2$ and $12 \rightarrow 3$, allowing to write [A.6] in an expanded form

$$\begin{bmatrix} E^H_{11} & E^H_{12} & E^H_{13} \\ E^H_{21} & E^H_{22} & E^H_{23} \\ E^H_{31} & E^H_{32} & E^H_{33} \end{bmatrix} = \begin{bmatrix} Q_{11} & Q_{12} & Q_{13} \\ Q_{21} & Q_{22} & Q_{23} \\ Q_{31} & Q_{32} & Q_{33} \end{bmatrix}, \quad \text{[A.7]}$$

where the terms Q_{ij}

$$Q_{ij} = \frac{1}{|Y|}\sum_{e=1}^{N} q_e^{(ij)}, \quad \text{[A.8]}$$

are the sums of element mutual energies $q_e^{(ij)}$

$$q_e^{(ij)} = (\mathbf{u}_e^{A(i)})^T\mathbf{k}_e\mathbf{u}_e^{A(j)}. \quad \text{[A.9]}$$

A.3. Periodic boundary conditions

The strain fields $\varepsilon_{pq}^{A(kl)}$ in [A.5] are evaluated by solving the base cell equilibrium problem subject to the unit test strains $\varepsilon_{pq}^{0(kl)}$. Under the assumption of periodicity, the displacement field of the base cell subjected to a given strain ε_{ij}^0 can be written as the sum of a macroscopic displacement field and a periodic fluctuation field u_i^* [MIC 99]

$$u_i = \varepsilon_{ij}^0 y_j + u_i^*. \quad \text{[A.10]}$$

In practice, equation [A.10] cannot be directly imposed on the boundaries because the periodic fluctuation term u_i^* is unknown. This general expression needs to be transformed into a certain number of explicit constraints between the corresponding pairs of nodes on the opposite surfaces of the base cell [XIA 03]. Consider a 2D base cell as shown in Figure A.5, the displacements on a pair of opposite boundaries are

$$\begin{cases} u_i^{k+} = \varepsilon_{ij}^0 y_j^{k+} + u_i^* \\ u_i^{k-} = \varepsilon_{ij}^0 y_j^{k-} + u_i^*, \end{cases} \qquad \text{[A.11]}$$

where superscripts "$k+$" and "$k-$" denote the pair of two opposite parallel boundary surfaces that are oriented perpendicular to the th direction ($k = 1, 2, 3$). The periodic term u_i^* can be eliminated through the difference between the displacements

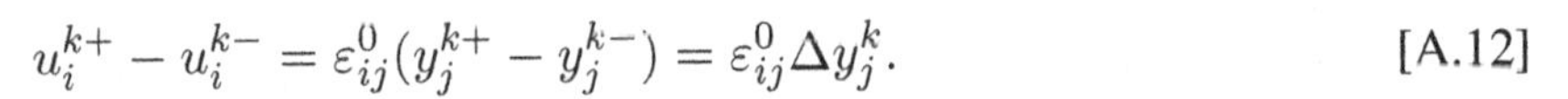

$$u_i^{k+} - u_i^{k-} = \varepsilon_{ij}^0 (y_j^{k+} - y_j^{k-}) = \varepsilon_{ij}^0 \Delta y_j^k. \qquad \text{[A.12]}$$

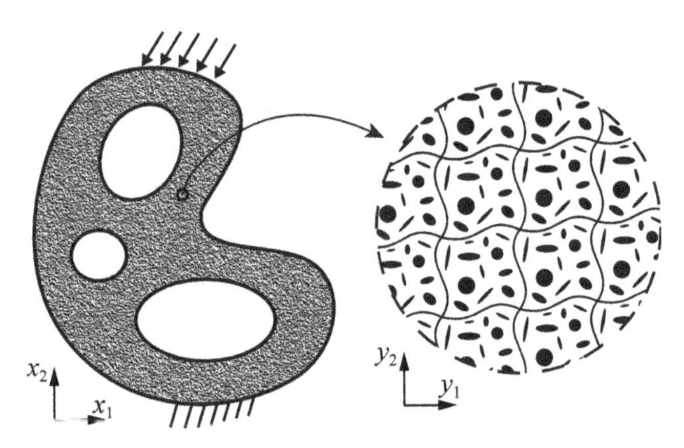

Figure A.4. *Illustration of a material point constituted by periodically patterned microstructures*

For any given parallelepiped base cell model, Δy_j^k is constant. In the case of Figure A.5, we have $\Delta y_1^1 = y_1^0, \Delta y_2^1 = 0$ and $\Delta y_1^2 = 0, \Delta y_2^2 = y_2^0$. Thus, with a specified ε_{ij}^0, the right-hand side of the equation becomes a constant

$$u_i^{k+} - u_i^{k-} = w_i^k, \qquad \text{[A.13]}$$

since $w_i^k = \varepsilon_{ij}^0 \Delta y_j^k$. This boundary condition can be directly imposed in the finite element model by constraining the corresponding pairs of nodal displacements. At the same time, this form of boundary conditions meets the periodicity and the continuity requirements for both displacement as well as stress when using displacement-based finite element analysis [XIA 06].

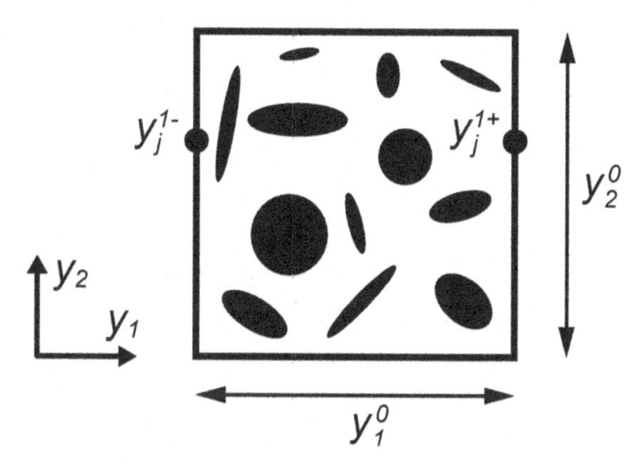

Figure A.5. *A 2D rectangular base cell model*

A.4. Optimization model

A.4.1. *Modified SIMP approach*

The base cell is discretized into N finite elements and the same number of density design variables $\boldsymbol{\rho} \in \mathbb{R}^N$ are correspondingly defined. Using the modified SIMP approach [SIG 07], the element Young's modulus E_e is defined as

$$E_e(\rho_e) = E_{\min} + \rho_e^p(E_0 - E_{\min}), \tag{A.14}$$

where E_0 is the Young's modulus of solid material and $E_{\min}$ is Young's modulus of the Ersatz material, which is an approximation for void material using compliant material [ALL 04b] to prevent the singularity of the stiffness matrix. ρ_e takes values between 0 and 1, with these limits corresponding to the Ersatz and solid materials, respectively. p is a penalization factor introduced to drive the density distribution closer toward the so-called black-and-white solution.

The mathematical formulation of the optimization problem reads as follows

$$\begin{aligned} \min_{\boldsymbol{\rho}} &: c(E^H_{ijkl}(\boldsymbol{\rho})) \\ \text{s.t.} &: \mathbf{K}\mathbf{U}^{A(kl)} = \mathbf{F}^{(kl)}, k, l = 1, \ldots, d \\ &: \textstyle\sum_{e=1}^{N} v_e \rho_e / |Y| \leq \vartheta \\ &: 0 \leq \rho_e \leq 1, e = 1, \ldots, N \end{aligned} \qquad \text{[A.15]}$$

where $\mathbf{K}$ is the global stiffness matrix, and $\mathbf{U}^{A(kl)}$ and $\mathbf{F}^{(kl)}$ are the global displacement vector and the external force vector of the test case (kl), respectively. d is the spatial dimension, v_e denotes the element volume and ϑ is the upper bound on the volume fraction. The objective $c(E^H_{ijkl})$ is a function of the homogenized stiffness tensors. For instance, in the 2D case, the maximization of the material bulk modulus corresponds to the minimization of

$$c = -\left(E_{1111} + E_{1122} + E_{2211} + E_{2222}\right), \qquad \text{[A.16]}$$

and the maximization of material shear modulus corresponds to the minimization of

$$c = -E_{1212}. \qquad \text{[A.17]}$$

A.4.2. *Numerical solution of the homogenization equations*

When both the geometry and the loading exhibit symmetries, which is the case here, the periodic boundary conditions presented in section A.3 can be simplified to conventional boundary conditions [HAS 98b]. To keep the derivations general, such simplification is not applied in the present work. Instead, the periodic boundary conditions are imposed in a direct manner (see section A.3). With regard to the finite element solution of equation [A.13], the direct solution scheme eliminating the redundant unknowns is adopted here. Note that, apart from the direct solution scheme, there exist two other types of solution schemes using penalty methods and Lagrange multipliers [MIC 99].

Separating the global displacement vector $\mathbf{U}$ into four parts: $\bar{\mathbf{U}}_1$ denotes the prescribed displacement values, $\mathbf{U}_2$ denotes the unknowns corresponding to the interior nodes, and $\mathbf{U}_3$ and $\mathbf{U}_4$ denote unknowns corresponding to the nodes located on the opposite boundaries of the base cell satisfying $\mathbf{U}_4 = \mathbf{U}_3 +$

$\bar{\mathbf{W}}$, where $\bar{\mathbf{W}}$ is a prescribed value computed via a given $\varepsilon^{0(kl)}$ according to equation [A.13]. The equilibrium equation in [A.15] can be expanded to

$$\begin{bmatrix} \mathbf{K}_{11} & \mathbf{K}_{12} & \mathbf{K}_{13} & \mathbf{K}_{14} \\ \mathbf{K}_{21} & \mathbf{K}_{22} & \mathbf{K}_{23} & \mathbf{K}_{24} \\ \mathbf{K}_{31} & \mathbf{K}_{32} & \mathbf{K}_{33} & \mathbf{K}_{34} \\ \mathbf{K}_{41} & \mathbf{K}_{42} & \mathbf{K}_{43} & \mathbf{K}_{44} \end{bmatrix} \begin{bmatrix} \bar{\mathbf{U}}_1 \\ \mathbf{U}_2 \\ \mathbf{U}_3 \\ \mathbf{U}_4 \end{bmatrix} = \begin{bmatrix} \mathbf{F}_1 \\ \mathbf{F}_2 \\ \mathbf{F}_3 \\ \mathbf{F}_4 \end{bmatrix}, \qquad \text{[A.18]}$$

where $\mathbf{F}_1$ is an unknown vector and is equal to the reaction forces at the nodes with prescribed displacements, $\mathbf{F}_2 = \mathbf{0}$ and $\mathbf{F}_3 + \mathbf{F}_4 = \mathbf{0}$, due to the periodicity assumption. Note that $\mathbf{K}$ is symmetric, i.e. $\mathbf{K}_{ij} = \mathbf{K}_{ji}$, in [A.18]. Eliminating the first row, adding the third and fourth rows, and using the relationship $\mathbf{U}_4 = \mathbf{U}_3 + \bar{\mathbf{W}}$, reduces equation [A.18] to

$$\begin{aligned} &\begin{bmatrix} \mathbf{K}_{22} & \mathbf{K}_{23} + \mathbf{K}_{24} \\ sym. & \mathbf{K}_{33} + \mathbf{K}_{34} + \mathbf{K}_{43} + \mathbf{K}_{44} \end{bmatrix} \begin{bmatrix} \mathbf{U}_2 \\ \mathbf{U}_3 \end{bmatrix} \\ &= - \begin{bmatrix} \mathbf{K}_{21} \\ \mathbf{K}_{31} + \mathbf{K}_{41} \end{bmatrix} \bar{\mathbf{U}}_1 - \begin{bmatrix} \mathbf{K}_{24} \\ \mathbf{K}_{34} + \mathbf{K}_{44} \end{bmatrix} \bar{\mathbf{W}}. \end{aligned} \qquad \text{[A.19]}$$

and allows for the solution of the system.

A.4.3. *Optimality criteria method*

Once the displacement solution is obtained, the optimization problem [A.15] is solved by means of a standard optimality criteria method. Following [BEN 03], the heuristic updating scheme is formulated as

$$\rho_e^{\text{new}} = \begin{cases} \max(0, \rho_e - m) & \text{if } \rho_e B_e^\eta \leq \max(0, \rho_e - m) \\ \min(1, \rho_e + m) & \text{if } \rho_e B_e^\eta \geq \min(1, \rho_e + m) \\ \rho_e B_e^\eta & \text{otherwise,} \end{cases} \qquad \text{[A.20]}$$

where m is a positive move limit, η is a numerical damping coefficient and B_e is obtained from the optimality condition as [BEN 03]

$$B_e = \frac{-\frac{\partial c}{\partial \rho_e}}{\lambda \frac{\partial V}{\partial \rho_e}}, \qquad \text{[A.21]}$$

where the Lagrange multiplier λ is chosen by means of a bisection algorithm to enforce the satisfaction of the constraint on material volume fraction. The sensitivity of the objective function $\partial c/\partial \rho_e$ is computed using the adjoint method [BEN 03]

$$\frac{\partial E^H_{ijkl}}{\partial \rho_e} = \frac{1}{|Y|} p \rho_e^{p-1} (E_0 - E_{\min}) (\mathbf{u}_e^{A(ij)})^T \mathbf{k}_0 \mathbf{u}_e^{A(kl)}, \qquad [A.22]$$

in accordance with the objective definition, where $\mathbf{k}_0$ is the element stiffness matrix for an element with unit Young's modulus. When a uniform mesh is used, the element volume v_e is set to 1 and therefore $\partial V/\partial \rho_e = 1$.

In order to ensure the existence of the solution to the optimization problem [A.15], sensitivity and density filtering schemes are used following [AND 11] to avoid the formation of checkerboard patterns and the mesh-dependent issues.

A.5. Matlab implementation

In this section, the Matlab code (see section A.8) is explained. The present code is built on top of the 88-line code [AND 11]. The first 38 lines are left unchanged. Material properties are defined in lines 4–6. The element stiffness matrix and the corresponding nodal informations are defined in lines 8–17. Matrices that are to be used for sensitivity and density filtering are predefined in lines 19–38. The design domain is assumed to be rectangular and discretized into square plane stress elements. The main program is called from the Matlab prompt by the command:

```
topX(nelx,nely,volfrac,penal,rmin,ft)
```

where `nelx` and `nely` denote the number of elements along the horizontal and vertical directions, respectively, `volfrac` is the prescribed volume fraction, `penal` is the penalization factor p, `rmin` is the filter radius and `ft` specifies whether sensitivity filtering (`ft=1`) or density filtering (`ft=2`) is to be used.

The following sections present the original parts of code developed in the scope of the current work. Apart from these, two minor changes are made to the original 88-line code: line 103, the stop condition is set to `1e-9` to enforce the satisfaction of the volume fraction constraint; line 111, `mean(xPhys(:))` is used for programming consistency.

A.5.1. *Lines 39–56: periodic boundary conditions*

Periodic boundary conditions (section A.3) are defined in lines 39–56. e0 defines the three unit test strain fields. The base cell shown in Figure A.6 is discretized into 3×3 elements for the purpose of illustration. The DOFs are divided into four sets as presented in equation [A.18]

$$\begin{cases} \mathtt{d1} = \{7,8,31,32,25,26,1,2\} \\ \mathtt{d15} = \{11,12,13,14,19,20,21,22\} \\ \mathtt{d3} = \{3,4,5,6,15,16,23,24\} \\ \mathtt{d4} = \{27,28,29,30,9,10,17,18\}, \end{cases} \qquad [A.23]$$

where d1 contains the DOFs of the four corner points (A, B, C, D), d3 contains the DOFs on the left and bottom boundaries except the corner DOFs, d4 contains the DOFs on the right and top boundaries except the corner DOFs, and d2 contains the remaining inner DOFs. In practice, one has to fix at least one node to avoid rigid body motion when solving the PBC problem. When point A is chosen to be fixed, points B, C, and D are prescribed with values corresponding to the three unit test strain fields computed according to equation [A.13] in lines 51–55. wfixed in line 56 corresponds to w_i^k in equation [A.13] and $\bar{\mathbf{W}}$ in equation [A.19], and is the constant difference vector between the DOFs of d3 and d4.

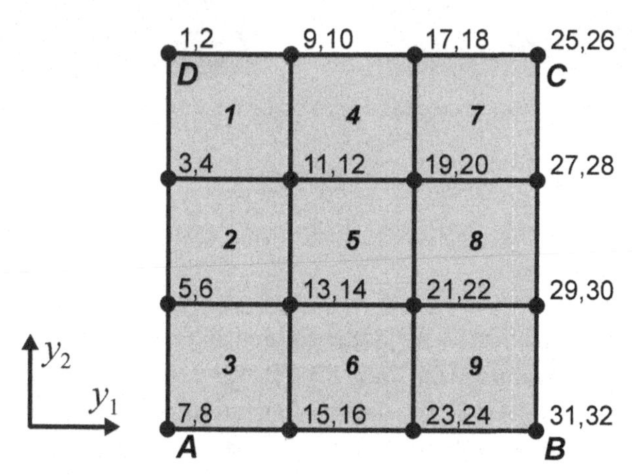

Figure A.6. *A base cell discretized into 3×3 elements*

A.5.2. *Lines 57–71: initialization*

In the 2D case, the base cell model needs to be evaluated three times corresponding to the three unit test strain fields. Lines 58–60 preallocate three cells `qe`, `Q` and `dQ` to store the element mutual energies, the summed mutual energies and the sensitivities of the summed mutual energies.

Lines 61–69 give the initial guess of material topology layout (Figure A.7, left). In structural compliance minimization designs [SIG 01], the initial guess usually consists of a uniformly distributed density field to avoid local minimum designs. However, this cannot be employed for material designs because the applied periodic boundary conditions would result in a uniformly distributed sensitivity field, thus making the variable update impossible. The influence of an initial guess on the final designs has been thoroughly discussed in [SIG 97, SIG 00] and [GIB 00]; however, the specific initial guesses are not provided. Following [AMS 10], we simply define a circular region with softer material at the center of the base cell as shown in Figure A.7.

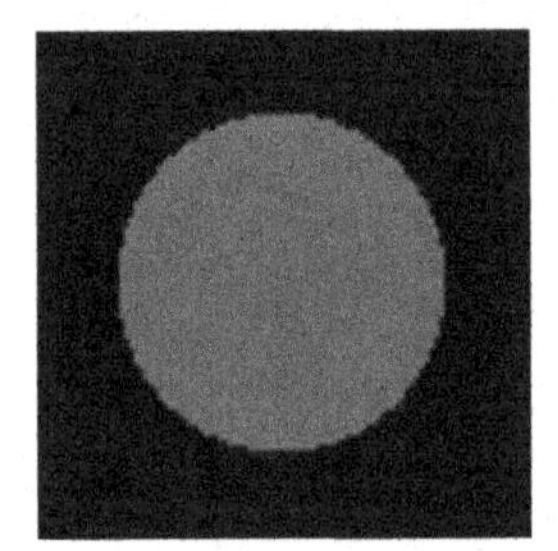
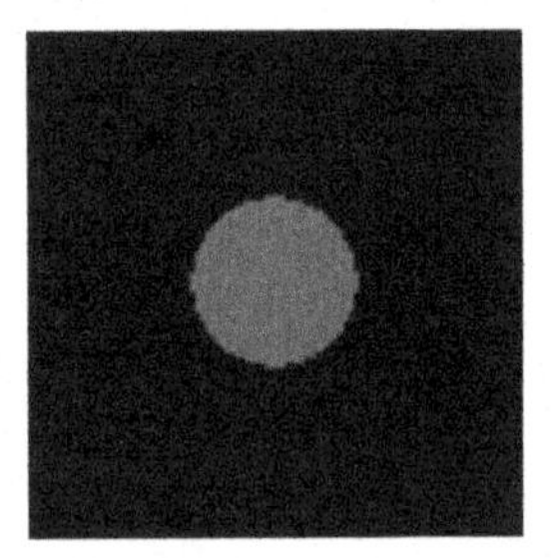

Figure A.7. *Two initial guess topologies for a* 100×100 *base cell*

A.5.3. *Lines 76–93: finite element solution, objective and sensitivity analysis*

The system in equation [A.19] is assembled and solved in lines 76–81. Cells `qe`, `Q` and `dQ` are evaluated in lines 83–90. Line 91 calculates the objective function in equation [A.16] for the maximization of material bulk modulus. Sensitivities are computed lines 92 and 93, and stored in `dc` and `dv`.

A.6. Illustrative examples

As discussed by Sigmund and Torquato [SIG 97], Sigmund [SIG 00] and Gibiansky and Sigmund [GIB 00], topology optimization design of materials with extreme properties allows for multiple local minima. The initial guess of material topology layout, the shape of base cell, filter radius, penalization factor and other parameters all have influence on the design solution. In the following examples, we show how to use the present Matlab code to design materials with extreme properties. All of the following tests are performed using Matlab 8.4.0.150421 (R2014b).

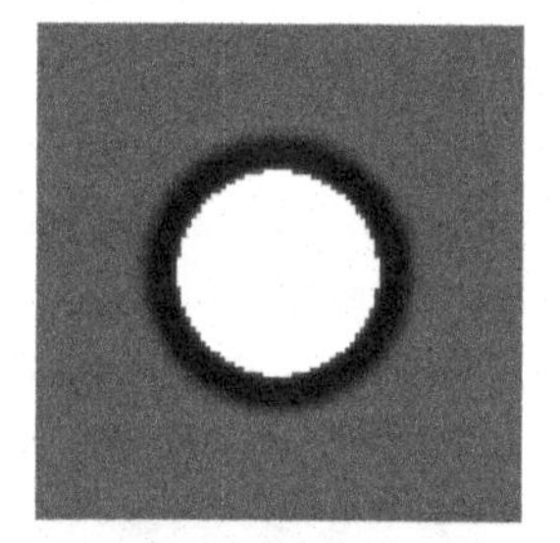

Figure A.8. *Materials with maximized bulk moduli obtained using sensitivity filtering (left, $c = -0.4388$, iteration 163) and density filtering (right, $c = -0.6537$, iteration 216) where penalty factor $p = 3$ and filter radius $r = 5$*

A.6.1. *Material bulk modulus maximization*

According to [BEN 03], the so-called one-length scale microstructures can be obtained by setting the filter radius to a comparatively large value, saying 10% of the cell length at the beginning iterations, then gradually decease its value during the optimization process. Here, we simply set the filter radius to 5. The penalization factor is set to 3. Materials with maximized bulk moduli shown in Figure A.8 can be obtained by calling:

```
topX(100,100,0.5,3,5,1), topX(100,100,0.5,3,5,2)
```

respectively. Both filtering schemes converge after around 200 iterations; however, the sensitivity filtering scheme fails in giving a clear structural layout.

To favor solutions with clearly discernible topologies, Bendsøe and Sigmund [BEN 03] proposed to gradually increase the penalization factor during the optimization process. Note that the topology may be driven closer toward a black–white solution if the penalization value is increased. It should be noted, however, that the problem in equation [A.15] is non-convex for values of $p > 1$. Thus, while high penalization values will result in cleaner topologies, the algorithm is more likely to get trapped in a local minimum.

In this work, the penalization factor is set as a constant value, thus we simply increase the value to 5. Recalling:

```
topX(100,100,0.5,5,5,1), topX(100,100,0.5,5,5,2)
```

respectively, we obtain materials shown in Figure A.9. It can be observed that the increased penalization value results in clearer topology layout in the case of sensitivity filtering, while it has little influence in the case of density filtering.

Figure A.9. *Materials with maximized bulk moduli obtained using sensitivity filtering (left, $c = -0.5636$, iteration 269) and density filtering (right, $c = -0.6207$, iteration 160) where penalty factor $p = 5$ and filter radius $r = 5$*

Similar to the implementation of penalization, filtering scheme is applied here to regularize the problem in equation [A.15], which means a smaller filter radius will result in a better solution because higher frequency details are allowed by the low-pass filer, thus enriching the solution. To show the influence of the filter radius on the solutions, the same problem is solved again with filter radius $r = 2$ by calling:

```
topX(100,100,0.5,5,2,1), topX(100,100,0.5,5,2,2)
```

respectively. As shown in Figure A.10, the decreased filter radius value results in a more detailed microstructure when using sensitivity filtering, while a slightly varied microstructure in the case of density filtering. As expected, both materials shown in Figure A.10 possess higher bulk moduli than those in Figure A.9 due to the decreased filter radius value.

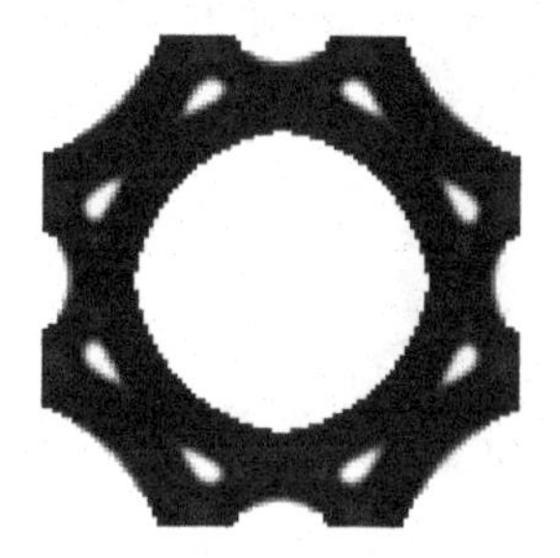 

Figure A.10. *Materials with maximized bulk moduli obtained using sensitivity filtering (left, $c = -0.6793$, iteration 183) and density filtering (right, $c = -0.6540$, iteration 129) where penalty factor $p = 5$ and filter radius $r = 2$*

We note that when using sensitivity filtering (Figure A.8, left), the minimum is obtained at iteration 10, after which the algorithm diverges. This is a common phenomenon for all the following tests when using sensitivity filtering. Sensitivity filtering scheme is fully heuristic in nature and was developed for the specific case of compliance minimization [SIG 01]. Thus, there is no guarantee that it will perform well for material designs. On the contrary, the density filtering scheme does not suffer from the concerned issue and has a more robust performance according to Figures A.8–A.10. Therefore, the density filtering scheme is in general preferable for the design of material microstructures.

The present Matlab code also allows the design of rectangular base cells. The two rectangular microstructures shown in Figure A.11 are obtained by calling:

```
topX(150,100,0.5,5,2,1),
topX(150,100,0.5,5,2,2)
```

using sensitivity filtering and density filtering, respectively. Again, it can be observed that a small value for the filter radius results in a more detailed microstructure in the case sensitivity filtering is used.

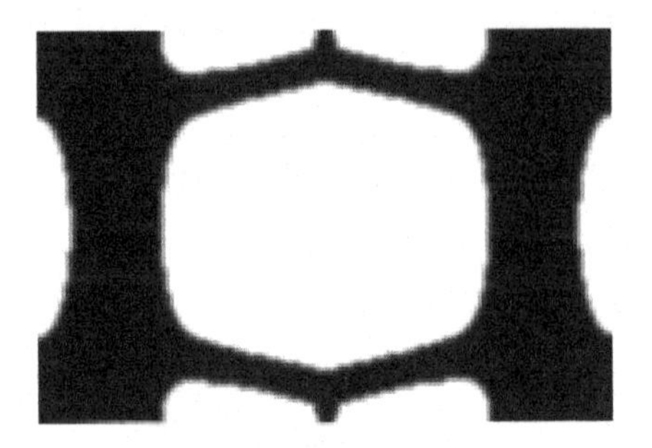

Figure A.11. *Materials with maximized bulk moduli obtained using sensitivity filtering (top, $c = -0.6730$, iteration 345) and density filtering (bottom, $c = -0.6317$, iteration 349) where penalty factor $p = 5$ and filter radius $r = 2$*

From these results, it can be argued that the results obtained using density filtering are less sensitive to the choice of optimization parameters such as penalization factor and filter radius.

A.6.2. *Material shear modulus maximization*

To design materials with maximized shear moduli, we may simply replace lines 91 and 92 of the code by:

```
c=-Q(3,3); dc=-dQ{3,3},
```

and the corresponding designs in Figure A.12 can be obtained by calling:

```
topX(100,100,0.5,3,5,1),
topX(100,100,0.5,3,5,2),
```

respectively. Compared to the maximization of material bulk modulus, material shear modulus maximization design is less sensitive to the choice of filtering scheme and does not require a high penalization factor to enforce topologically clear designs.

 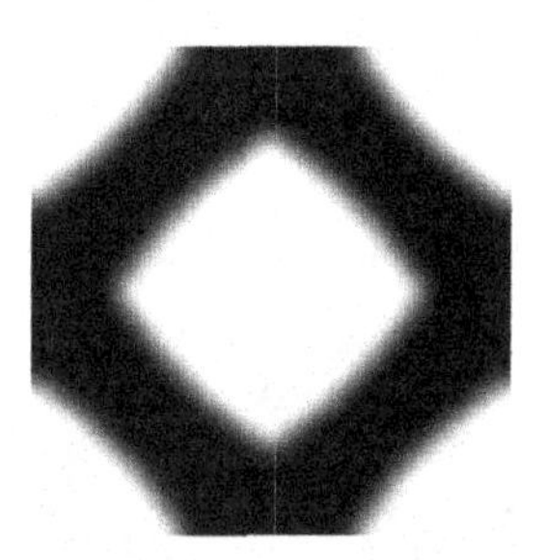

Figure A.12. *Materials with maximized shear moduli obtained using sensitivity filtering (left, $c = -0.1256$, iteration 12) and density filtering (right, $c = -0.1213$, iteration 72) where penalty factor $p = 3$ and filter radius $r = 5$*

Replacing `min(nelx,nely)/3` by `min(nelx,nely)/6` in line 64, we consider another initial guess with a smaller circular region at the center of the base cell (Figure A.7, right). With the modified initial guess, the same function calls result in different material topology designs as shown in Figure A.13. The difference between the results in Figures A.12 and A.13 indicates that the choice of initial material topology guess has a severe influence on the final designs. As argued by Bendsøe and Sigmund [BEN 03], different initial guesses may lead to different microstructures that may possess similar material properties due to the non-uniqueness of the solution. Sometimes the microstructural topologies obtained from different initial guesses are in fact the shifted versions of the same topology. It is also suggested to start with an old design to a similar problem, which may save considerable amount of computing time as can be seen from a recent work by Andreassen *et al.* [AND 14b] for 3D material designs with negative Poisson ratio.

By decreasing the convergence criterion from 0.01 to 0.001, i.e. replacing line 73 by:

```
while (change > 0.001)
```

materials with higher shear moduli as shown in Figure A.14 can be achieved by the same function calls as above while requiring more design iterations.

A discussion on this topology transition, given in a recent review paper [SIG 13], stated that "the optimization rapidly finds a fairly good design but requires a very large number of iterations for just slight improvements in objective function but rather large changes in geometry" and "to the best of our knowledge remedies for this issues are unknown and we pose it as a challenge to the community to come up with more efficient updates for continuous variable approaches".

Figure A.13. *Materials with maximized shear moduli obtained from a modified initial guess using sensitivity filtering (left, $c = -0.1118$, iteration 58) and density filtering (right, $c = -0.1043$, iteration 64) where penalty factor $p = 3$ and filter radius $r = 5$*

Figure A.14. *Materials with maximized shear moduli obtained from the modified initial guess using sensitivity filtering (left, $c = -0.1175$, iteration 179) and density filtering (right, $c = -0.1057$, iteration 734) where the convergence criterion set as 0.001, penalty factor $p = 3$ and filter radius $r = 5$*

A.6.3. *Materials with negative Poisson's ratio*

The design of materials with negative Poisson's ratio $\mu = \frac{E_{1122}}{E_{1111}}$ using topology optimization is a challenging subject. As shown by Sigmund [SIG 94], the construction of negative Poisson's ratio materials with the present model using an OC-type optimizer is difficult. Successful design of negative Poisson's ratio materials requires imposing additional constraints, for instance on isotropy and on bulk modulus. In order to consider multiple constraints in the design, we can either use a specially developed OC-type method [YIN 01], or employ more versatile mathematical programing optimizers such as the method of moving asymptotes [SVA 87] implemented by Bendsøe and Sigmund [BEN 03] and more recently by Andreassen *et al.* [AND 14b] and Wang *et al.* [WAN 14c].

In order to construct negative Poisson's ratio materials with the present model, we propose to define a relaxed form of objective function

$$c = E_{1122} - \beta^l (E_{1111} + E_{2222}), \qquad \text{[A.24]}$$

where $\beta \in (0,1)$ is a fixed parameter defined by the user and exponential l is the design iteration number. With this objective function, an optimizer tends to maximize material horizontal and vertical stiffness moduli at the beginning iterations. When the optimization process advances, i.e. when l increases, an optimizer tends to minimize the value of E_{1122} such that materials with negative Poisson's ratios are constructed.

Choosing $\beta = 0.8$, we need to replace lines 91 and 92 of the Matlab code by:

```
c = Q(1,2)-(0.8loop){*}(Q(1,1)+Q(2,2));
dc = dQ{1,2}-(0.8loop){*}(dQ{1,1}+dQ{2,2}),
```

and modify line 105 to:

```
xnew = max(0,max(x-move,min(1,...
min(x+move,x.{*}(-dc./dv/lmid))))),
```

omitting the numerical damping coefficient. The damping coefficient has to be removed here because both positive and negative sensitivities appear when the objective function of [A.24] is considered. The constraint on the material volume fraction may not be active during the optimization process as can be seen from the following tests. Design solutions are more sensitive to the choice of the initial guess and other parameters as compared to the previous two cases.

As the numerical damping coefficient is removed, the move limit is decreased from 0.2 to 0.1 to stabilize the algorithm, i.e. modifying line 102 to:

```
l1=0;
l2=1e9; move = 0.1
```

With all these modifications to the Matlab code, calling:

```
topX(100,100,0.5,3,5,1)}, topX(100,100,0.5,3,5,2)
```

results in the materials with negative Poisson's ratio as shown Figure A.15. Note that neither result of the two shown in Figure A.15 is a converged solution. When employing sensitivity filtering scheme, the design process oscillates between two topologies after around 40 iterations and after around 30 iterations in the case of density filtering. The volume fraction constraint is not active in the case of density filtering, which implies that the solution reached a stable state where any addition of material results in a worse topology ($volume\,fraction = 0.437$). Theoretically speaking, the Lagrange multiplier λ in equation [A.21] should be equal to zero within this context. A zero-valued Lagrange multiplier would trigger a division by zero error in the OC update scheme. This is fortunately avoided because of an unintentional particularity of the bisection algorithm used to find this Lagrange multiplier: the lowest value it could ever take is $\lambda_{\min} \approx (10^{-9} - 0)/2 = 5 \times 10^{-10}$, thus allowing the algorithm to continue.

As discussed in section A.6.2, we consider an alternative initial guess (Figure A.7, right). Substituting `min(nelx,nely)/6` in line 64, the same function calls result in the designs shown in Figure A.16. The left figure in Figure A.16 is one of the two states when oscillation begins after around 50 iterations when using sensitivity filtering. The right figure in Figure A.16 with Poisson's ratio $\mu = -0.446$ is a converged solution at iteration 113 when using density filtering, where we note that the volume fraction constraint is satisfied. It is shown by this example that the OC-type optimizer is able to generate well-designed negative Poisson's ratio materials without defining additional constraints provided that the tuning parameters are carefully adjusted.

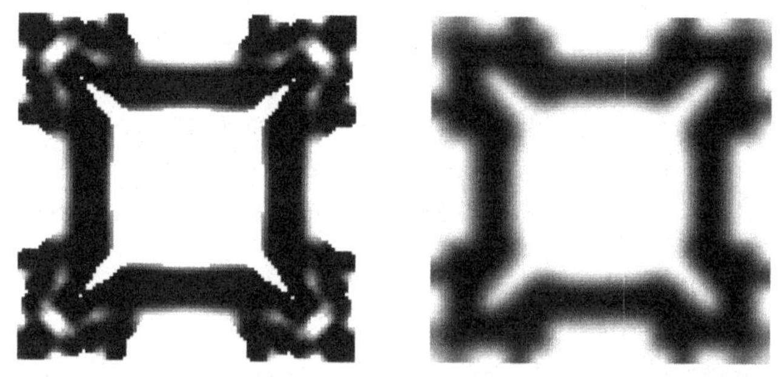

Figure A.15. *Materials with negative Poisson's ratio obtained using sensitivity filtering (left, $\mu = -0.346$, volume fraction $= 0.5$, not converged) and density filtering (right, $\mu = -0.239$, volume fraction $= 0.437$, not converged) where penalty factor $p = 3$ and filter radius $r = 5$*

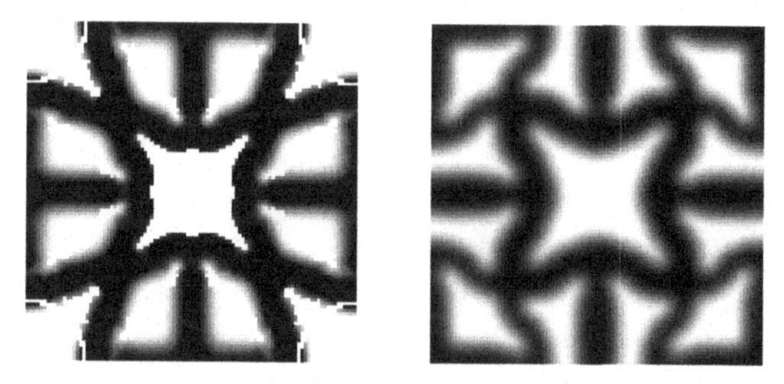

Figure A.16. *Materials with negative Poisson's ratio obtained from a modified initial guess using sensitivity filtering (left, $\mu = -0.323$, volume fraction $= 0.5$, not converged) and density filtering (right, $\mu = -0.448$, volume fraction $= 0.5$, iteration 113) where penalty factor $p = 3$ and filter radius $r = 5$*

A.7. Concluding remarks

This paper extends the 88-line code [AND 11] to the design of materials with extreme properties. The adoption of an energy-based homogenization approach instead of the asymptotic approach significantly simplifies the numerical implementation. Periodic boundary conditions and the elimination of the redundant unknowns are presented in detail together with the corresponding numerical implementation.

The present code uses an OC-type optimizer with a single constraint on volume fraction and is able to design materials with extreme bulk and shear modulus. With a proposed relaxed objective function [A.24], the present code also allows to construct materials with negative Poisson's ratio without introducing additional constraints such as symmetry or on isotropy.

Note that the discussions on the influence of the optimization parameters and the filtering schemes on the solutions in the present work are mostly based on pure observations. More rigorous explanations would require further investigations.

A.8. Matlab Code "topX.m"

```
1  %% PERIODIC MATERIAL MICROSTRUCTURE DESIGN
2  function topX(nelx,nely,volfrac,penal,rmin,ft)
3  %% MATERIAL PROPERTIES
4  E0 = 1;
5  Emin = 1e-9;
6  nu = 0.3;
7  %% PREPARE FINITE ELEMENT ANALYSIS
8  A11 = [12 3 -6 -3; 3 12 3  0; -6 3 12 -3; -3 0 -3 12];
9  A12 = [-6 -3 0 3; -3 -6 -3 -6; 0 -3 -6 3; 3 -6 3 -6];
10 B11 = [-4 3 -2 9;  3 -4 -9 4; -2 -9 -4 -3; 9 4 -3 -4];
11 B12 = [ 2 -3 4 -9; -3 2 9 -2; 4  9  2  3; -9 -2 3  2];
12 KE = 1/(1-nu^2)/24*([A11 A12;A12' A11]+nu*[B11 B12;B12'
       B11]);
13 nodenrs = reshape(1:(1+nelx)*(1+nely),1+nely,1+nelx);
14 edofVec = reshape(2*nodenrs(1:end-1,1:end-1)+1,nelx*nely
       ,1);
15 edofMat = repmat(edofVec,1,8)+repmat([0 1 2*nely+[2 3 0
       1] -2 -1],nelx*nely,1);
16 iK = reshape(kron(edofMat,ones(8,1))',64*nelx*nely,1);
17 jK = reshape(kron(edofMat,ones(1,8))',64*nelx*nely,1);
18 %% PREPARE FILTER
19 iH = ones(nelx*nely*(2*(ceil(rmin)-1)+1)^2,1);
20 jH = ones(size(iH));
21 sH = zeros(size(iH));
22 k = 0;
23 for i1 = 1:nelx
```

```
24    for j1 = 1:nely
25      e1 = (i1-1)*nely+j1;
26      for i2 = max(i1-(ceil(rmin)-1),1):min(i1+(ceil(rmin)
            -1),nelx)
27        for j2 = max(j1-(ceil(rmin)-1),1):min(j1+(ceil(rmin
              )-1),nely)
28          e2 = (i2-1)*nely+j2;
29          k = k+1;
30          iH(k) = e1;
31          jH(k) = e2;
32          sH(k) = max(0,rmin-sqrt((i1-i2)^2+(j1-j2)^2));
33        end
34      end
35    end
36  end
37  H = sparse(iH,jH,sH);
38  Hs = sum(H,2);
39  %% PERIODIC BOUNDARY CONDITIONS
40  e0 = eye(3);
41  ufixed = zeros(8,3);
42  U = zeros(2*(nely+1)*(nelx+1),3);
43  alldofs = (1:2*(nely+1)*(nelx+1));
44  n1 = [nodenrs(end,[1,end]),nodenrs(1,[end,1])];
45  d1 = reshape([(2*n1-1);2*n1],1,8);
46  n3 = [nodenrs(2:end-1,1)',nodenrs(end,2:end-1)];
47  d3 = reshape([(2*n3-1);2*n3],1,2*(nelx+nely-2));
48  n4 = [nodenrs(2:end-1,end)',nodenrs(1,2:end-1)];
49  d4 = reshape([(2*n4-1);2*n4],1,2*(nelx+nely-2));
50  d2 = setdiff(alldofs,[d1,d3,d4]);
51  for j = 1:3
52    ufixed(3:4,j) =[e0(1,j),e0(3,j)/2;e0(3,j)/2,e0(2,j)]*[
          nelx;0];
53    ufixed(7:8,j) = [e0(1,j),e0(3,j)/2;e0(3,j)/2,e0(2,j)
          ]*[0;nely];
54    ufixed(5:6,j) = ufixed(3:4,j)+ufixed(7:8,j);
55  end
56  wfixed = [repmat(ufixed(3:4,:),nely-1,1); repmat(ufixed
        (7:8,:),nelx-1,1)];
```

```
57 %% INITIALIZE ITERATION
58 qe = cell(3,3);
59 Q = zeros(3,3);
60 dQ = cell(3,3);
61 x = repmat(volfrac,nely,nelx);
62 for i = 1:nelx
63   for j = 1:nely
64     if sqrt((i-nelx/2-0.5)^2+(j-nely/2-0.5)^2) < min(
         nelx,nely)/3
65       x(j,i) = volfrac/2;
66     end
67   end
68 end
69 xPhys = x;
70 change = 1;
71 loop = 0;
72 %% START ITERATION
73 while (change > 0.01)
74   loop = loop+1;
75   %% FE-ANALYSIS
76   sK = reshape(KE(:)*(Emin+xPhys(:)'.^penal*(E0-Emin))
       ,64*nelx*nely,1);
77   K = sparse(iK,jK,sK); K = (K+K')/2;
78   Kr = [K(d2,d2), K(d2,d3)+K(d2,d4); K(d3,d2)+K(d4,d2),
       K(d3,d3)+K(d4,d3)+K(d3,d4)+K(d4,d4)];
79   U(d1,:) = ufixed;
80   U([d2,d3],:) = Kr\(-[K(d2,d1); K(d3,d1)+K(d4,d1)]*
       ufixed-[K(d2,d4); K(d3,d4)+K(d4,d4)]*wfixed);
81   U(d4,:) = U(d3,:)+wfixed;
82   %% OBJECTIVE FUNCTION AND SENSITIVITY ANALYSIS
83   for i = 1:3
84     for j = 1:3
85       U1 = U(:,i); U2 = U(:,j);
86       qe{i,j} = reshape(sum((U1(edofMat)*KE).*U2(edofMat)
           ,2),nely,nelx)/(nelx*nely);
87       Q(i,j) = sum(sum((Emin+xPhys.^penal*(E0-Emin)).*qe{
           i,j}));
```

```
88        dQ{i,j} = penal*(E0-Emin)*xPhys.^(penal-1).*qe{i,j
              };
89      end
90    end
91    c = -(Q(1,1)+Q(2,2)+Q(1,2)+Q(2,1));
92    dc = -(dQ{1,1}+dQ{2,2}+dQ{1,2}+dQ{2,1});
93    dv = ones(nely,nelx);
94    %% FILTERING/MODIFICATION OF SENSITIVITIES
95    if ft == 1
96      dc(:) = H*(x(:).*dc(:))./Hs./max(1e-3,x(:));
97    elseif ft == 2
98      dc(:) = H*(dc(:)./Hs);
99      dv(:) = H*(dv(:)./Hs);
100   end
101   %% OPTIMALITY CRITERIA UPDATE OF DESIGN VARIABLES AND
          PHYSICAL DENSITIES
102   l1 = 0; l2 = 1e9; move = 0.2;
103   while (l2-l1 > 1e-9)
104     lmid = 0.5*(l2+l1);
105     xnew = max(0,max(x-move,min(1,min(x+move,x.*sqrt(-dc
            ./dv/lmid)))));
106     if ft == 1
107       xPhys = xnew;
108     elseif ft == 2
109       xPhys(:) = (H*xnew(:))./Hs;
110     end
111     if mean(xPhys(:)) > volfrac, l1 = lmid; else l2 =
            lmid; end
112   end
113   change = max(abs(xnew(:)-x(:)));
114   x = xnew;
115   %% PRINT RESULTS
116   fprintf(' It.:%5i Obj.:%11.4f Vol.:%7.3f ch.:%7.3f\n',
          loop,c, mean(xPhys(:)),change);
117   %% PLOT DENSITIES
118   colormap(gray); imagesc(1-xPhys); caxis([0 1]); axis
          equal; axis off; drawnow;
119 end
```

Bibliography

[ALE 15] ALEXANDERSEN J., LAZAROV B.S., "Topology optimisation of manufacturable microstructural details without length scale separation using a spectral coarse basis preconditioner", *Computer Methods in Applied Mechanics and Engineering*, vol. 290, pp. 156–182, 2015.

[ALL 04a] ALLAIRE G., JOUVE F., MAILLOT H., "Topology optimization for minimum stress design with the homogenization method", *Structural and Multidisciplinary Optimization*, vol. 28, nos. 2–3, pp. 87–98, 2004.

[ALL 04b] ALLAIRE G., JOUVE F., TOADER A.M., "Structural optimization using sensitivity analysis and a level-set method", *Journal of Computational Physics*, vol. 194, no. 1, pp. 363–393, 2004.

[ALM 10] ALMEIDA S.R.M., PAULINO G.H., SILVA E.C.N., "Layout and material gradation in topology optimization of functionally graded structures: a global-local approach", *Structural and Multidisciplinary Optimization*, vol. 42, no. 6, pp. 855–868, 2010.

[AMS 10] AMSTUTZ S., GIUSTI S., NOVOTNY A., *et al.*, "Topological derivative for multi-scale linear elasticity models applied to the synthesis of microstructures", *International Journal for Numerical Methods in Engineering*, vol. 84, no. 6, pp. 733–756, 2010.

[AND 11] ANDREASSEN E., CLAUSEN A., SCHEVENELS M., *et al.*, "Efficient topology optimization in MATLAB using 88 lines of code", *Structural and Multidisciplinary Optimization*, vol. 43, no. 1, pp. 1–16, 2011.

[AND 14a] ANDREASSEN E., JENSEN J., "Topology optimization of periodic microstructures for enhanced dynamic properties of viscoelastic composite materials", *Structural and Multidisciplinary Optimization*, vol. 49, no. 5, pp. 695–705, 2014.

[AND 14b] ANDREASSEN E., LAZAROV B., SIGMUND O., "Design of manufacturable 3D extremal elastic microstructure", *Mechanics of Materials*, vol. 69, pp. 1–10, 2014.

[BAD 10] BADER B., KOLDA T., MATLAB Tensor Toolbox Version 2.4k, 2010, Available at http://www.sandia.gov/tgkolda/TensorToolbox/.

[BEL 03] BELYTSCHKO T., XIAO S., PARIMI C., "Topology optimization with implicit functions and regularization", *International Journal for Numerical Methods in Engineering*, vol. 57, no. 8, pp. 1177–1196, 2003.

[BEN 88] BENDSØE M.P., KIKUCHI N., "Generating optimal topologies in structural design using a homogenization method", *Computer Methods in Applied Mechanics and Engineering*, vol. 71, no. 2, pp. 197–224, 1988.

[BEN 89] BENDSØE M.P., "Optimal shape design as a material distribution problem", *Structural Optimization*, vol. 1, no. 4, pp. 193–202, 1989.

[BEN 95] BENDSØE M.P., DIAZ A.R., LIPTON R., *et al.*, "Optimal design of material properties and material distribution for multiple loading conditions", *International Journal for Numerical Methods in Engineering*, vol. 38, no. 7, pp. 1149–1170, 1995.

[BEN 96] BENDSØE M., GUEDES J., PLAXTON S., *et al.*, "Optimization of structure and material properties for solids composed of softening material", *International Journal of Solids and Structures*, vol. 33, no. 12, pp. 1799–1813, 1996.

[BEN 99] BENDSØE M.P., SIGMUND O., "Material interpolation schemes in topology optimization", *Archive of Applied Mechanics*, vol. 69, pp. 635–654, 1999.

[BEN 03] BENDSØE M.P., SIGMUND O., *Topology Optimization: Theory, Methods and Applications*, Springer-Verlag, Berlin, 2003.

[BOG 12] BOGOMOLNY M., AMIR O., "Conceptual design of reinforced concrete structures using topology optimization with elastoplastic material modeling", *International Journal for Numerical Methods in Engineering*, vol. 90, no. 13, pp. 1578–1597, 2012.

[BOU 03] BOURDIN B., CHAMBOLLE A., "Design-dependent loads in topology optimization", *ESAIM – Control, Optimisation and Calculus of Variations*, no. 9, pp. 19–48, 2003.

[BRE 98] BREITKOPF P., TOUZOT G., VILLON P., "Consistency approach and diffuse derivation in element free methods based on moving least squares approximation", *Computer Assisted Mechanics and Engineering Sciences*, vol. 5, no. 4, pp. 479–501, 1998.

[BRE 04] BREITKOPF P., RASSINEUX A., SAVIGNAT J.M.,*et al.*, "Integration constraint in diffuse element method", *Computer Methods in Applied Mechanics and Engineering*, vol. 193, no. 12–14, pp. 1203–1220, 2004.

[BRU 03] BRUNS T., TORTORELLI D., "An element removal and reintroduction strategy for the topology optimization of structures and compliant mechanisms", *International Journal for Numerical Methods in Engineering*, vol. 57, no. 10, pp. 1413–1430, 2003.

[BRU 12] BRUGGI M., DUYSINX P., "Topology optimization for minimum weight with compliance and stress constraints", *Structural and Multidisciplinary Optimization*, vol. 46, no. 3, pp. 369–384, 2012.

[BUH 00] BUHL T., PEDERSEN C., SIGMUND O., "Stiffness design of geometrically nonlinear structures using topology optimization", *Structural and Multidisciplinary Optimization*, vol. 19, no. 2, pp. 93–104, 2000.

[BUR 04] BURGER M., HACKL B., RING W., "Incorporating topological derivatives into level set methods", *Journal of Computational Physics*, vol. 194, no. 1, pp. 344–362, 2004.

[CAD 13] CADMAN J., ZHOU S., CHEN Y., *et al.*, "On design of multi-functional microstructural materials", *Journal of Materials Science*, vol. 48, no. 1, pp. 51–66, 2013.

[CAI 14] CAI S., ZHANG W., ZHU J., *et al.*, "Stress constrained shape and topology optimization with fixed mesh: a B-spline finite cell method combined with level set function", *Computer Methods in Applied Mechanics and Engineering*, vol. 278, pp. 361–387, 2014.

[CAI 15] CAI S., ZHANG W., "Stress constrained topology optimization with free-form design domains", *Computer Methods in Applied Mechanics and Engineering*, vol. 289, pp. 267–290, 2015.

[CAR 70] CARROLL J., CHANG J.J., "Analysis of individual differences in multidimensional scaling via an n-way generalization of "Eckart-Young" decomposition", *Psychometrika*, vol. 35, no. 3, pp. 283–319, 1970.

[CHA 08] CHALLIS V.J., ROBERTS A.P., WILKINS A.H., "Design of three dimensional isotropic microstructures for maximized stiffness and conductivity", *International Journal of Solids and Structures*, vol. 45, nos. 14–15, pp. 4130–4146, 2008.

[CHA 10] CHALLIS V.J., "A discrete level-set topology optimization code written in Matlab", *Structural and Multidisciplinary Optimization*, vol. 41, no. 3, pp. 453–464, 2010.

[CHA 12] CHALLIS V.J., GUEST J.K., GROTOWSKI J.F., *et al.*, "Computationally generated cross-property bounds for stiffness and fluid permeability using topology optimization", *International Journal of Solids and Structures*, vol. 49, nos. 23–24, pp. 3397–3408, 2012.

[CHE 14] CHEN W., LIU S., "Topology optimization of microstructures of viscoelastic damping materials for a prescribed shear modulus", *Structural and Multidisciplinary Optimization*, vol. 50, no. 2, pp. 287–296, 2014.

[CHO 03] CHO S., JUNG H.-S., "Design sensitivity analysis and topology optimization of displacement-loaded non-linear structures", *Computer Methods in Applied Mechanics and Engineering*, vol. 192, nos. 22–23, pp. 2539–2553, 2003.

[CLÉ 12] CLÉMENT A., SOIZE C., YVONNET J., "Computational nonlinear stochastic homogenization using a nonconcurrent multiscale approach for hyperelastic heterogeneous microstructures analysis", *International Journal for Numerical Methods in Engineering*, vol. 91, no. 8, pp. 799–824, 2012.

[CLÉ 13] CLÉMENT A., SOIZE C., YVONNET J., "Uncertainty quantification in computational stochastic multiscale analysis of nonlinear elastic materials", *Computer Methods in Applied Mechanics and Engineering*, vol. 254, pp. 61–82, 2013.

[COE 08] COELHO P.G., FERNANDES P.R., GUEDES J.M., *et al.*, "A hierarchical model for concurrent material and topology optimisation of three-dimensional structures", *Structural and Multidisciplinary Optimization*, vol. 35, no. 2, pp. 107–115, 2008.

[COE 12] COENEN E.W.C., KOUZNETSOVA V.G., GEERS M.G.D., "Multi-scale continuous-discontinuous framework for computational-homogenization-localization", *Journal of the Mechanics and Physics of Solids*, vol. 60, no. 8, pp. 1486–1507, 2012.

[COE 15] COELHO P.G., GUEDES J.M., RODRIGUES H.C., "Multiscale topology optimization of bi-material laminated composite structures", *Composite Structures*, vol. 132, pp. 495–505, 2015.

[CRE 13] CREMONESI M., NÉRON D., GUIDAULT P.A., LADEVÈZE P., "A PGD-based homogenization technique for the resolution of nonlinear multiscale problems", *Computer Methods in Applied Mechanics and Engineering*, vol. 267, pp. 275–292, 2013.

[DEA 14] DEATON J.D., GRANDHI R.V., "A survey of structural and multidisciplinary continuum topology optimization: post 2000", *Structural and Multidisciplinary Optimization*, vol. 49, no. 1, pp. 1–38, 2014.

[DEN 13] DENG J., YAN J., CHENG G., "Multi-objective concurrent topology optimization of thermoelastic structures composed of homogeneous porous material", *Structural and Multidisciplinary Optimization*, vol. 47, no. 4, pp. 583–597, 2013.

[DUV 84] DUVA J., HUTCHINSON J., "Constitutive potentials for dilutely voided nonlinear materials", *Mechanics of Materials*, vol. 3, no. 1, pp. 41–54, 1984.

[DUY 98] DUYSINX P., BENDSØE M.P., "Topology optimization of continuum structures with local stress constraints", *International Journal for Numerical Methods in Engineering*, vol. 43, no. 8, pp. 1453–1478, 1998.

[DVO 92] DVORAK G., BENVENISTE Y., "On transformation strains and uniform fields in multiphase elastic media", *Proceedings of the Royal Society of London A*, vol. 437, no. 437, pp. 291–310, 1992.

[DVO 94] DVORAK G., BAHEI-EL-DIN Y., WAFA A., "The modeling of inelastic composite materials with the transformation field analysis", *Modelling and Simulation in Material Science and Engineering*, vol. 2, no. 2, pp. 571–586, 1994.

[ELH 13] EL HALABI F., GONZÁLEZ D., CHICO A., DOBLARÉ M., "FE^2 multiscale in linear elasticity based on parametrized microscale models using proper generalized decomposition", *Computer Methods in Applied Mechanics and Engineering*, vol. 257, pp. 183–202, 2013.

[ESC 94] ESCHENAUER H., KOBELEV V., SCHUMACHER A., "Bubble method for topology and shape optimization of structures", *Structural Optimization*, vol. 8, no. 1, pp. 42–51, 1994.

[FEY 00] FEYEL F., CHABOCHE J., "FE^2 multiscale approach for modelling the elastoviscoplastic behaviour of long fibre SiC/Ti composite materials", *Computer Methods in Applied Mechanics and Engineering*, vol. 183, nos. 3–4, pp. 309–330, 2000.

[FOR 09] FORRESTER A., KEANE A., "Recent advances in surrogate-based optimization", *Progress in Aerospace Sciences*, vol. 45, no. 1–3, pp. 50–79, 2009.

[FRI 10] FRITZEN F., BÖHLKE T., "Three-dimensional finite element implementation of the nonuniform transformation field analysis", *International Journal for Numerical Methods in Engineering*, vol. 84, no. 7, pp. 803–829, 2010.

[FRI 11] FRITZEN F., BÖHLKE T., "Nonuniform transformation field analysis of materials with morphological anisotropy", *Composites Science and Technology*, vol. 71, pp. 433–442, 2011.

[FRI 13a] FRITZEN F., BÖHLKE T., "Reduced basis homogenization of viscoelastic composites", *Composites Science and Technology*, vol. 76, pp. 84–91, 2013.

[FRI 13b] FRITZEN F., LEUSCHNER M., "Reduced basis hybrid computational homogenization based on a mixed incremental formulation", *Computer Methods in Applied Mechanics and Engineering*, vol. 260, pp. 143–154, 2013.

[FRI 14] FRITZEN F., HODAPP M., LEUSCHNER M., "GPU accelerated computational homogenization based on a variational approach in a reduced basis framework", *Computer Methods in Applied Mechanics and Engineering*, vol. 278, pp. 186–217, 2014.

[FRI 15a] FRITZEN F., HODAPP M., "The Finite Element Square Reduced (FE^{2R}) method with GPU acceleration: towards three-dimensional two-scale simulations", *International Journal for Numerical Methods in Engineering*, 2015, DOI: 10.1002/nme.5188.

[FRI 15b] FRITZEN F., XIA L., LEUSCHNER M., *et al.*, "Topology optimization of multiscale elastoviscoplastic structures", *International Journal for Numerical Methods in Engineering*, 2015, DOI: 10.1002/nme.5122.

[FUJ 01] FUJII D., CHEN B.C., KIKUCHI N., "Composite material design of two-dimensional structures using the homogenization design method", *International Journal for Numerical Methods in Engineering*, vol. 50, no. 9, pp. 2031–2051, 2001.

[FUL 10] FULLWOOD D.T., NIEZGODA S.R., ADAMS B.L.,*et al.*, "Microstructure sensitive design for performance optimization", *Progress in Materials Science*, vol. 55, no. 6, pp. 477–562, 2010.

[GAN 07] GANAPATHYSUBRAMANIAN B., ZABARAS N., "Modeling diffusion in random heterogeneous media: data-driven models, stochastic collocation and the variational multiscale method", *Journal of Computational Physics*, vol. 226, no. 1, pp. 326–353, 2007.

[GAO 12] GAO T., ZHANG W., DUYSINX P., "A bi-value coding parameterization scheme for the discrete optimal orientation design of the composite laminate", *International Journal for Numerical Methods in Engineering*, vol. 91, no. 1, pp. 98–114, 2012.

[GAO 13] GAO T., ZHANG W.H., DUYSINX P., "Simultaneous design of structural layout and discrete fiber orientation using bi-value coding parameterization and volume constraint", *Structural and Multidisciplinary Optimization*, vol. 48, no. 6, pp. 1075–1088, 2013.

[GEA 01] GEA H., LUO J., "Topology optimization of structures with geometrical nonlinearities", *Computers and Structures*, vol. 79, nos. 20–21, pp. 1977–1985, 2001.

[GEE 10] GEERS M.G.D., KOUZNETSOVA V.G., BREKELMANS W.A.M., "Multi-scale computational homogenization: trends and challenges", *Journal of Computational and Applied Mathematics*, vol. 234, no. 7, pp. 2175–2182, 2010.

[GHO 01] GHOSH S., LEE K., RAGHAVAN P., "A multi-level computational model for multi-scale damage analysis in composite and porous materials", *International Journal of Solids and Structures*, vol. 38, no. 14, pp. 2335–2385, 2001.

[GIB 00] GIBIANSKY L., SIGMUND O., "Multiphase composites with extremal bulk modulus", *Journal of the Mechanics and Physics of Solids*, vol. 48, no. 3, pp. 461–498, 2000.

[GU 12] GU X., ZHU J., ZHANG W., "The lattice structure configuration design for stereolithography investment casting pattern using topology optimization", *Rapid Prototyping Journal*, vol. 18, no. 5, pp. 353–361, 2012.

[GUE 90] GUEDES J., KIKUCHI N., "Preprocessing and postprocessing for materials based on the homogenization method with adaptive finite element methods", Computer Methods in Applied Mechanics and Engineering, vol. 83, no. 2, pp. 143–198, 1990.

[GUE 06] GUEST J.K., PRÉVOST J.H., "Optimizing multifunctional materials: design of microstructures for maximized stiffness and fluid permeability", *International Journal of Solids and Structures*, vol. 43, no. 22–23, pp. 7028–7047, 2006.

[GUE 07] GUEST J.K., PRÉVOST J.H., "Design of maximum permeability material structures", *Computer Methods in Applied Mechanics and Engineering*, vol. 196, nos. 4–6, pp. 1006–1017, 2007.

[GUE 08] GUESSASMA S., BABIN P., DELLA VALLE G., *et al.*, "Relating cellular structure of open solid food foams to their Young's modulus: finite element calculation", *International Journal of Solids and Structures*, vol. 45, no. 10, pp. 2881–2896, 2008.

[GUO 11] GUO X., ZHANG W., WANG M., *et al.*, "Stress-related topology optimization via level set approach", *Computer Methods in Applied Mechanics and Engineering*, vol. 200, nos. 47–48, pp. 3439–3452, 2011.

[GUO 15] GUO X., ZHAO X., ZHANG W., *et al.*, "Multi-scale robust design and optimization considering load uncertainties", *Computer Methods in Applied Mechanics and Engineering*, vol. 283, pp. 994–1009, 2015.

[HAL 75] HALPHEN B., NGUYEN Q.S., "Sur les matériaux standards generalisés", *Journal de Mécanique*, vol. 14, no. 1, pp. 39–63, 1975.

[HAS 83] HASHIN Z., "Analysis of composite materials – a survey", *Journal of Applied Mechanics, Transactions ASME*, vol. 50, no. 3, pp. 481–505, 1983.

[HAS 98a] HASSANI B., HINTON E., "A review of homogenization and topology optimization I – homogenization theory for media with periodic structure", *Computers and Structures*, vol. 69, no. 6, pp. 707–717, 1998.

[HAS 98b] HASSANI B., HINTON E., "A review of homogenization and topology opimization II – analytical and numerical solution of homogenization equations", *Computers and Structures*, vol. 69, no. 6, pp. 719–738, 1998.

[HER 14] HERNANDEZ J., OLIVER J., HUESPE A., *et al.*, "High-performance model reduction techniques in computational multiscale homogenization", *Computer Methods in Applied Mechanics and Engineering*, vol. 276, pp. 149–189, 2014.

[HIL 52] HILL R., "The elastic behaviour of a crystalline aggregate", *Proceedings of the Physical Society. Section A*, vol. 65, no. 5, pp. 349–354, 1952.

[HIL 63] HILL R., "Elastic properties of reinforced solids: some theoretical principles", *Journal of the Mechanics and Physics of Solids*, vol. 11, no. 5, pp. 357–372, 1963.

[HUA 07a] HUANG X., XIE Y., "Convergent and mesh-independent solutions for the bi-directional evolutionary structural optimization method", *Finite Elements in Analysis and Design*, vol. 43, no. 14, pp. 1039–1049, 2007.

[HUA 07b] HUANG X., XIE Y., LU G., "Topology optimization of energy-absorbing structures", *International Journal of Crashworthiness*, vol. 12, no. 6, pp. 663–675, 2007.

[HUA 08] HUANG X., XIE Y.M., "Topology optimization of nonlinear structures under displacement loading", *Engineering Structures*, vol. 30, no. 7, pp. 2057–2068, 2008.

[HUA 09] HUANG X., XIE Y.M., "Bi-directional evolutionary topology optimization of continuum structures with one or multiple materials", *Computational Mechanics*, vol. 43, no. 3, pp. 393–401, 2009.

[HUA 10a] HUANG X., XIE Y.M., "A further review of ESO type methods for topology optimization", *Structural and Multidisciplinary Optimization*, vol. 41, no. 5, pp. 671–683, 2010.

[HUA 10b] HUANG X., XIE Y.M., *Topology Optimization of Continuum Structures: Methods and Applications*, John Wiley & Sons, Chichester, 2010.

[HUA 11] HUANG X., RADMAN A., XIE Y.M., "Topological design of microstructures of cellular materials for maximum bulk or shear modulus", *Computational Materials Science*, vol. 50, no. 6, pp. 1861–1870, 2011.

[HUA 12] HUANG X., XIE Y.M., JIA B., *et al.*, "Evolutionary topology optimization of periodic composites for extremal magnetic permeability and electrical permittivity", *Structural and Multidisciplinary Optimization*, vol. 46, no. 3, pp. 385–398, 2012.

[HUA 13] HUANG X., ZHOU S.W., XIE Y.M., *et al.*, "Topology optimization of microstructures of cellular materials and composites for macrostructures", *Computational Materials Science*, vol. 67, pp. 397–407, 2013.

[HUA 15] HUANG X., ZHOU S., SUN G., LI G., *et al.*, "Topology optimization for microstructures of viscoelastic composite materials", *Computer Methods in Applied Mechanics and Engineering*, vol. 283, pp. 503–516, 2015.

[IBR 03] IBRAHIMBEGOVIC A., MARKOVIC D., "Strong coupling methods in multi-phase and multi-scale modeling of inelastic behavior of heterogeneous structures", *Computer Methods in Applied Mechanics and Engineering*, vol. 192, nos. 28-30, pp. 3089–3107, 2003.

[IBR 10] IBRAHIMBEGOVIC A., PAPADRAKAKIS M., "Multi-scale models and mathematical aspects in solid and fluid mechanics", *Computer Methods in Applied Mechanics and Engineering*, vol. 199, nos. 21–22, p. 1241, 2010.

[JUN 04] JUNG D., GEA H., "Topology optimization of nonlinear structures", *Finite Elements in Analysis and Design*, vol. 40, no. 11, pp. 1417–1427, 2004.

[KAT 14] KATO J., YACHI D., TERADA K., *et al.*, "Topology optimization of micro-structure for composites applying a decoupling multi-scale analysis", *Structural and Multidisciplinary Optimization*, vol. 49, no. 4, pp. 595–608, 2014.

[KAT 15] KATO J., HOSHIBA H., TAKASE S., *et al.*, "Analytical sensitivity in topology optimization for elastoplastic composites", *Structural and Multidisciplinary Optimization*, vol. 52, no. 3, pp. 507–526, 2015.

[KIE 10] KIERS H., "Towards a standardized notation and terminology in multiway analysis", *Journal of Chemometrics*, vol. 14, no. 3, pp. 105–122, 2010.

[KOH 86] KOHN R., STRANG G., "Optimal design and relaxation of variational problems (part I)", *Commun Pure Applied Math*, vol. 39, no. 1, 1986.

[KOU 01] KOUZNETSOVA V., BREKELMANS W.A.M., BAAIJENS F.P.T., "An approach to micro-macro modeling of heterogeneous materials", *Computational Mechanics*, vol. 27, no. 1, pp. 37–48, 2001.

[LAM 10] LAMARI H., AMMAR A., CARTRAUD P., *et al.*, "Routes for efficient computational homogenization of nonlinear materials using the proper generalized decompositions", *Archives of Computational Methods in Engineering*, vol. 17, no. 4, pp. 373–391, 2010.

[LAN 81] LANCASTER P., SALKAUSKAS K., "Surfaces generated by moving least squares methods", *Mathematics of Computation*, vol. 87, pp. 141–158, 1981.

[LE 10] LE C., NORATO J., BRUNS T., *et al.*, "Stress-based topology optimization for continua", *Structural and Multidisciplinary Optimization*, vol. 41, no. 4, pp. 605–620, 2010.

[LE 15] LE B., YVONNET J., HE Q.-C., "Computational homogenization of nonlinear elastic materials using Neural Networks", *International Journal for Numerical Methods in Engineering*, vol. 104, no. 12, pp. 1061–1084, 2015.

[LI 01] LI Q., STEVEN G., XIE Y., "A simple checkerboard suppression algorithm for evolutionary structural optimization", *Structural and Multidisciplinary Optimization*, vol. 22, no. 3, pp. 230–239, 2001.

[LIU 05] LIU Z., KORVINK J., HUANG R., "Structure topology optimization: fully coupled level set method via FEMLAB", *Structural and Multidisciplinary Optimization*, vol. 29, no. 6, pp. 407–417, 2005.

[LIU 12] LIU S., HOU Y., SUN X., *et al.*, "A two-step optimization scheme for maximum stiffness design of laminated plates based on lamination parameters", *Composite Structures*, vol. 94, no. 12, pp. 3529–3537, 2012.

[LUO 08] LUO J., LUO Z., CHEN L., *et al.*, "A semi-implicit level set method for structural shape and topology optimization", *Journal of Computational Physics*, vol. 227, no. 11, pp. 5561–5581, 2008.

[LUO 15] LUO Y., WANG M., KANG Z., "Topology optimization of geometrically nonlinear structures based on an additive hyperelasticity technique", *Computer Methods in Applied Mechanics and Engineering*, vol. 286, pp. 422–441, 2015.

[LV 14] LV J., ZHANG H., CHEN B., "Shape and topology optimization for closed liquid cell materials using extended multiscale finite element method", *Structural and Multidisciplinary Optimization*, vol. 49, no. 3, pp. 367–385, 2014.

[MAH 06] MAHDAVI A., BALAJI R., FRECKER M., *et al.*, "Topology optimization of 2D continua for minimum compliance using parallel computing", *Structural and Multidisciplinary Optimization*, vol. 32, no. 2, pp. 121–132, 2006.

[MAU 98] MAUTE K., SCHWARZ S., RAMM E., "Adaptive topology optimization of elastoplastic structures", *Structural Optimization*, vol. 15, no. 2, pp. 81–91, 1998.

[MIC 99] MICHEL J.C., MOULINEC H., SUQUET P., "Effective properties of composite materials with periodic microstructure: a computational approach", *Computer Methods in Applied Mechanics and Engineering*, vol. 172, no. 1–4, pp. 109–143, 1999.

[MIC 03] MICHEL J.C., SUQUET P., "Nonuniform transformation field analysis", *International Journal of Solids and Structures*, vol. 40, pp. 6937–6955, 2003.

[MIC 04] MICHEL J.C., SUQUET P., "Computational analysis of nonlinear composite structures using the nonuniform transformation field analysis", *Computer Methods in Applied Mechanics and Engineering*, vol. 193, pp. 5477–5502, 2004.

[MIE 99] MIEHE C., SCHRÖDER J., SCHOTTE J., "Computational homogenization analysis in finite plasticity simulation of texture development in polycrystalline materials", *Computer Methods in Applied Mechanics and Engineering*, vol. 171, nos. 3–4, pp. 387–418, 1999.

[MIE 02] MIEHE C., "Strain-driven homogenization of inelastic microstructures and composites based on an incremental variational formulation", *International Journal for Numerical Methods in Engineering*, vol. 55, pp. 1285–1322, 2002.

[MIL 13] MILED B., RYCKELYNCK D., CANTOURNET S., "A priori hyper-reduction method for coupled viscoelastic-viscoplastic composites", *Computers and Structures*, vol. 119, pp. 95–103, 2013.

[MOS 14] MOSBY M., MATOUS K., "Hierarchically parallel coupled finite strain multiscale solver for modeling heterogeneous layers", *International Journal for Numerical Methods in Engineering*, vol. 102, nos. 3–4, pp. 748–765, 2014.

[NAK 13] NAKSHATRALA P.B., TORTORELLI D.A., NAKSHATRALA K.B., "Nonlinear structural design using multiscale topology optimization. Part I: static formulation", *Computer Methods in Applied Mechanics and Engineering*, vol. 261–262, pp. 167–176, 2013.

[NAY 92] NAYROLES B., TOUZOT G., VILLON P., "Generalizing the finite element method: diffuse approximation and diffuse elements", *Computational Mechanics*, vol. 10, no. 5, pp. 307–318, 1992.

[NEV 00] NEVES M.M., RODRIGUES H., GUEDES J.M., "Optimal design of periodic linear elastic microstructures", *Computers and Structures*, vol. 76, no. 1, pp. 421–429, 2000.

[NEV 02] NEVES M.M., SIGMUND O., BENDSØE M.P., "Topology optimization of periodic microstructures with a penalization of highly localized buckling modes", *International Journal for Numerical Methods in Engineering*, vol. 54, no. 6, pp. 809–834, 2002.

[NIU 09] NIU B., YAN J., CHENG G., "Optimum structure with homogeneous optimum cellular material for maximum fundamental frequency", *Structural and Multidisciplinary Optimization*, vol. 39, no. 2, pp. 115–132, 2009.

[NIX 98] NIX W., GAO H., "Indentation size effects in crystalline materials: a law for strain gradient plasticity", *Journal of the Mechanics and Physics of Solids*, vol. 46, no. 3, pp. 411–425, 1998.

[OSK 07] OSKAY C., FISH J., "Eigendeformation-based reduced order homogenization for failure analysis of heterogeneous materials", *Computer Methods in Applied Mechanics and Engineering*, vol. 196, no. 7, pp. 1216–1243, 2007.

[PAU 09] PAULINO G.H., SILVA E.C.N., LE C.H., "Optimal design of periodic functionally graded composites with prescribed properties", *Structural and Multidisciplinary Optimization*, vol. 38, no. 5, pp. 469–489, 2009.

[PED 01] PEDERSEN C., BUHL T., SIGMUND O., "Topology synthesis of large-displacement compliant mechanisms", *International Journal for Numerical Methods in Engineering*, vol. 50, no. 12, pp. 2683–2705, 2001.

[QUE 00] QUERIN O.M., YOUNG V., STEVEN G.P., *et al.*, "Computational efficiency and validation of bi-directional evolutionary structural optimization", *Computer Methods in Applied Mechanics and Engineering*, vol. 189, no. 2, pp. 559–573, 2000.

[QUE 05] QUEIPO N.B., HAFTKA R.T., SHYY W., *et al.*, "Surrogate-based analysis and optimization", *Progress in Aerospace Sciences*, vol. 41, no. 1, pp. 1–28, 2005.

[RAG 13] RAGHAVAN B., XIA L., BREITKOPF P., *et al.*, "Towards simultaneous reduction of both input and output spaces for interactive simulation-based structural design", *Computer Methods in Applied Mechanics and Engineering*, vol. 265, no. 1, pp. 174–185, 2013.

[ROD 02] RODRIGUES H., GUEDES J.M., BENDSØE M.P., "Hierarchical optimization of material and structure", *Structural and Multidisciplinary Optimization*, vol. 24, no. 1, pp. 1–10, 2002.

[ROZ 01] ROZVANY G., "Aims, scope, methods, history and unified terminology of computer-aided topology optimization in structural mechanics", *Structural and Multidisciplinary Optimization*, vol. 21, no. 2, pp. 90–108, 2001.

[SCH 01] SCHWARZ S., MAUTE K., RAMM E., "Topology and shape optimization for elastoplastic structural response", *Computer Methods in Applied Mechanics and Engineering*, vol. 190, no. 15–17, pp. 2135–2155, 2001.

[SET 00] SETHIAN J.A., WIEGMANN A., "Structural boundary design via level set and immersed interface methods", *Journal of Computational Physics*, vol. 163, no. 2, pp. 489–528, 2000.

[SET 05] SETOODEH S., ABDALLA M.M., GÜRDAL Z., "Combined topology and fiber path design of composite layers using cellular automata", *Structural and Multidisciplinary Optimization*, vol. 30, no. 6, pp. 413–421, 2005.

[SET 06] SETOODEH S., ABDALLA M., GÜRDAL Z., "Design of variable-stiffness laminates using lamination parameters", *Composites Part B: Engineering*, vol. 37, no. 4–5, pp. 301–309, 2006.

[SIG 94] SIGMUND O., "Materials with prescribed constitutive parameters: an inverse homogenization problem", *International Journal of Solids and Structures*, vol. 31, no. 17, pp. 2313–2329, 1994.

[SIG 97] SIGMUND O., TORQUATO S., "Design of materials with extreme thermal expansion using a three-phase topology optimization method", *Journal of the Mechanics and Physics of Solids*, vol. 45, no. 6, pp. 1037–1067, 1997.

[SIG 00] SIGMUND O., "New class of extremal composites", *Journal of the Mechanics and Physics of Solids*, vol. 48, no. 2, pp. 397–428, 2000.

[SIG 01] SIGMUND O., "A 99 line topology optimization code written in matlab", *Structural and Multidisciplinary Optimization*, vol. 21, no. 2, pp. 120–127, 2001.

[SIG 07] SIGMUND O., "Morphology–based black and white filters for topology optimization", *Structural and Multidisciplinary Optimization*, vol. 33, nos. 4–5, pp. 401–424, 2007.

[SIG 13] SIGMUND O., MAUTE K., "Topology optimization approaches – a comparative review", *Structural and Multidisciplinary Optimization*, vol. 48, no. 6, pp. 1031–1055, 2013.

[SMI 98] SMIT R.J.M., BREKELMANS W.A.M., MEIJER H.E.H., "Prediction of the mechanical behavior of nonlinear heterogeneous systems by multi-level finite element modeling", *Computer Methods in Applied Mechanics and Engineering*, vol. 155, nos. 1–2, pp. 181–192, 1998.

[SOK 99] SOKOLOWSKI J., ZOCHOWSKI A., "On the topological derivative in shape optimization", *SIAM Journal on Control and Optimization*, vol. 37, no. 4, pp. 1251–1272, 1999.

[SU 10] SU W., LIU S., "Size-dependent optimal microstructure design based on couple-stress theory", *Structural and Multidisciplinary Optimization*, vol. 42, no. 2, pp. 243–254, 2010.

[SUR 10] SURESH K., "A 199-line Matlab code for pareto-optimal tracing in topology optimization", *Structural and Multidisciplinary Optimization*, vol. 42, no. 5, pp. 665–679, 2010.

[SUZ 91] SUZUKI K., KIKUCHI N., "A homogenization method for shape and topology optimization", *Computer Methods in Applied Mechanics and Engineering*, vol. 93, no. 3, pp. 291–318, 1991.

[SVA 87] SVANBERG K., "Method of moving asymptotes – a new method for structural optimization", *International Journal for Numerical Methods in Engineering*, vol. 24, no. 2, pp. 359–373, 1987.

[TAL 12a] TALISCHI C., PAULINO G., PEREIRA A., *et al.*, "PolyTop: a Matlab implementation of a general topology optimization framework using unstructured polygonal finite element meshes", *Structural and Multidisciplinary Optimization*, vol. 45, no. 3, pp. 329–357, 2012.

[TAL 12b] TALISCHI C., PAULINO G., PEREIRA A., *et al.*, "PolyMesher: a general-purpose mesh generator for polygonal elements written in Matlab", *Structural and Multidisciplinary Optimization*, vol. 45, no. 3, pp. 309–328, 2012.

[TAN 02] TANSKANEN P., "The evolutionary structural optimization method: theoretical aspects", *Computer Methods in Applied Mechanics and Engineering*, vol. 191, nos. 47–48, pp. 5485–5498, 2002.

[TEM 07a] TEMIZER I., WRIGGERS P., "An adaptive method for homogenization in orthotropic nonlinear elasticity", *Computer Methods in Applied Mechanics and Engineering*, vol. 196, no. 35-36, pp. 3409–3423, 2007.

[TEM 07b] TEMIZER I., ZOHDI T., "A numerical method for homogenization in non-linear elasticity", *Computational Mechanics*, vol. 40, no. 2, pp. 281–298, 2007.

[THE 99] THEOCARIS P.S., STAVROULAKI G.E., "Optimal material design in composites: an iterative approach based on homogenized cells", *Computer Methods in Applied Mechanics and Engineering*, vol. 169, nos. 1–2, pp. 31–42, 1999.

[TRA 11] TRAN A., YVONNET J., HE Q.-C., *et al.*, "A simple computational homogenization method for structures made of linear heterogeneous viscoelastic materials", *Computer Methods in Applied Mechanics and Engineering*, vol. 200, nos. 45–46, pp. 2956–2970, 2011.

[VAN 13] VAN DIJK N., MAUTE K., LANGELAAR M., *et al.*, "Level-set methods for structural topology optimization: a review", *Structural and Multidisciplinary Optimization*, vol. 48, no. 3, pp. 437–472, 2013.

[WAN 03] WANG M.Y., WANG X., GUO D., "A level set method for structural topology optimization", *Computer Methods in Applied Mechanics and Engineering*, vol. 192, nos. 1–2, pp. 227–246, 2003.

[WAN 14a] WANG F., LAZAROV B., SIGMUND O., *et al.*, "Interpolation scheme for fictitious domain techniques and topology optimization of finite strain elastic problems", *Computer Methods in Applied Mechanics and Engineering*, vol. 276, pp. 453–472, 2014.

[WAN 14b] WANG F., SIGMUND O., JENSEN J., "Design of materials with prescribed nonlinear properties", *Journal of the Mechanics and Physics of Solids*, vol. 69, no. 1, pp. 156–174, 2014.

[WAN 14c] WANG Y., LUO Z., ZHANG N., *et al.*, "Topological shape optimization of microstructural metamaterials using a level set method", *Computational Materials Science*, vol. 87, pp. 178–186, 2014.

[WEI 10] WEI P., WANG M., XING X., "A study on X-FEM in continuum structural optimization using a level set model", *CAD Computer Aided Design*, vol. 42, no. 8, pp. 708–719, 2010.

[XIA 03] XIA Z., ZHANG Y., ELLYIN F., "A unified periodical boundary conditions for representative volume elements of composites and applications", *International Journal of Solids and Structures*, vol. 40, no. 8, pp. 1907–1921, 2003.

[XIA 06] XIA Z., ZHOU C., YONG Q., *et al.*, "On selection of repeated unit cell model and application of unified periodic boundary conditions in micro-mechanical analysis of composites", *International Journal of Solids and Structures*, vol. 43, no. 2, pp. 266–278, 2006.

[XIA 10] XIAO M., BREITKOPF P., FILOMENO COELHO R., *et al.*, "Model reduction by CPOD and Kriging: application to the shape optimization of an intake port", *Structural and Multidisciplinary Optimization*, vol. 41, no. 4, pp. 555–574, 2010.

[XIA 12] XIA L., ZHU J., ZHANG W., "A superelement formulation for the efficient layout design of complex multi-component system", *Structural and Multidisciplinary Optimization*, vol. 45, no. 5, pp. 643–655, 2012.

[XIA 13a] XIA L., RAGHAVAN B., BREITKOPF P., *et al.*, "Numerical material representation using proper orthogonal decomposition and diffuse approximation", *Applied Mathematics and Computation*, vol. 224, pp. 450–462, 2013.

[XIA 13b] XIA L., ZHU J., ZHANG W., *et al.*, "An implicit model for the integrated optimization of component layout and structure topology", *Computer Methods in Applied Mechanics and Engineering*, vol. 257, pp. 87–102, 2013.

[XIA 14a] XIA L., BREITKOPF P., "Concurrent topology optimization design of material and structure within FE^2 nonlinear multiscale analysis framework", *Computer Methods in Applied Mechanics and Engineering*, vol. 278, pp. 524–542, 2014.

[XIA 14b] XIA L., BREITKOPF P., "A reduced multiscale model for nonlinear structural topology optimization", *Computer Methods in Applied Mechanics and Engineering*, vol. 280, pp. 117–134, 2014.

[XIA 15a] XIA L., BREITKOPF P., "Design of of materials using topology optimization and energy-based homogenization approach in Matlab", *Structural and Multidisciplinary Optimization*, 2015.

[XIA 15b] XIA L., BREITKOPF P., "Multiscale structural topology optimization with an approximate constitutive model for local material microstructure", *Computer Methods in Applied Mechanics and Engineering*, vol. 286, pp. 147–167, 2015.

[XIA 15c] XIA L., BREITKOPF P., "Recent advances on topology optimization of multiscale nonlinear structures", submitted to *Archives of Computational Methods in Engineering*.

[XIE 93] XIE Y.M., STEVEN G.P., "A simple evolutionary procedure for structural optimization", *Computers and Structures*, vol. 49, no. 5, pp. 885–896, 1993.

[XIE 97] XIE Y.M., STEVEN G.P., *Evolutionary Structural Optimization*, Springer-Verlag, London, 1997.

[XU 11] XU Y., ZHANG W., "Numerical modelling of oxidized microstructure and degraded properties of 2D C/SiC composites in air oxidizing environments below 800 °C", *Materials Science and Engineering A*, vol. 528, no. 27, pp. 7974–7982, 2011.

[XU 12] XU Y., ZHANG W., "A strain energy model for the prediction of the effective coefficient of thermal expansion of composite materials", *Computational Materials Science*, vol. 53, no. 1, pp. 241–250, 2012.

[XU 15] XU B., XIE Y., "Concurrent design of composite macrostructure and cellular microstructure under random excitations", *Composite Structures*, vol. 123, pp. 65–77, 2015.

[YAN 14] YAN X., HUANG X., ZHA Y., *et al.*, "Concurrent topology optimization of structures and their composite microstructures", *Computers and Structures*, vol. 133, pp. 103–110, 2014.

[YI 00] YI Y.-M., PARK S.-H., YOUN S.-K., "Design of microstructures of viscoelastic composites for optimal damping characteristics", *International Journal of Solids and Structures*, vol. 37, no. 35, pp. 4791–4810, 2000.

[YIN 01] YIN L., YANG W., "Optimality criteria method for topology optimization under multiple constraints", *Computers and Structures*, vol. 79, nos. 20–21, pp. 1839–1850, 2001.

[YOO 05] YOON G., KIM Y., "Element connectivity parameterization for topology optimization of geometrically nonlinear structures", *International Journal of Solids and Structures*, vol. 42, no. 7, pp. 1983–2009, 2005.

[YOO 07] YOON G., KIM Y., "Topology optimization of material-nonlinear continuum structures by the element connectivity parameterization", *International Journal for Numerical Methods in Engineering*, vol. 69, no. 10, pp. 2196–2218, 2007.

[YUA 09] YUAN Z., FISH J., "Multiple scale eigendeformation-based reduced order homogenization", *Computer Methods in Applied Mechanics and Engineering*, vol. 198, nos. 21–26, pp. 2016–2038, 2009.

[YUG 95] YUGE K., KIKUCHI N., "Optimization of a frame structure subjected to a plastic deformation", *Structural Optimization*, vol. 10, no. 3–4, pp. 197–208, 1995.

[YUG 99] YUGE K., IWAI N., KIKUCHI N., "Optimization of 2-D structures subjected to nonlinear deformations using the homogenization method", *Structural Optimization*, vol. 17, no. 4, pp. 286–299, 1999.

[YVO 07] YVONNET J., HE Q.C., "The reduced model multiscale method (R3M) for the non-linear homogenization of hyperelastic media at finite strains", *Journal of Computational Physics*, vol. 223, no. 1, pp. 341–368, 2007.

[YVO 09] YVONNET J., GONZALEZ D., HE Q.C., "Numerically explicit potentials for the homogenization of nonlinear elastic heterogeneous materials", *Computer Methods in Applied Mechanics and Engineering*, vol. 198, nos. 33–36, pp. 2723–2737, 2009.

[YVO 13] YVONNET J., MONTEIRO E., HE Q.C., "Computational homogenization method and reduced database model for hyperelastic heterogeneous structures", *International Journal for Multiscale Computational Engineering*, vol. 11, no. 3, pp. 201–225, 2013.

[ZHA 02] ZHANG T., GOLUB G.H., "Rank-one approximation to high order tensors", *SIAM Journal on Matrix Analysis and Applications*, vol. 23, no. 2, pp. 534–550, 2002.

[ZHA 06] ZHANG W., SUN S., "Scale-related topology optimization of cellular materials and structures", *International Journal for Numerical Methods in Engineering*, vol. 68, no. 9, pp. 993–1011, 2006.

[ZHA 07] ZHANG W., DAI G., WANG F., *et al.*, "Using strain energy-based prediction of effective elastic properties in topology optimization of material microstructures", *Acta Mechanica Sinica/Lixue Xuebao*, vol. 23, no. 1, pp. 77–89, 2007.

[ZHA 13] ZHANG W., GUO X., WANG M., *et al.*, "Optimal topology design of continuum structures with stress concentration alleviation via level set method", *International Journal for Numerical Methods in Engineering*, vol. 93, no. 9, pp. 942–959, 2013.

[ZHO 91] ZHOU M., ROZVANY G., "The COC algorithm, Part II: topological, geometrical and generalized shape optimization", *Computer Methods in Applied Mechanics and Engineering*, vol. 89, nos. 1–3, pp. 309–336, 1991.

[ZHO 01] ZHOU M., ROZVANY G., "On the validity of ESO type methods in topology optimization", *Structural and Multidisciplinary Optimization*, vol. 21, no. 1, pp. 80–83, 2001.

[ZHU 07] ZHU J., ZHANG W., QIU K., "Bi-directional evolutionary topology optimization using element replaceable method", *Computational Mechanics*, vol. 40, no. 1, pp. 97–109, 2007.

[ZHU 15] ZHU J., ZHANG W., XIA L., "Topology optimization in aircraft and aerospace structures design", *Archives of Computational Methods in Engineering*, 2015, DOI 10.1007/s11831-015-9151-2.

[ZUO 13] ZUO Z., HUANG X., RONG J., *et al.*, "Multi-scale design of composite materials and structures for maximum natural frequencies", *Materials and Design*, vol. 51, pp. 1023–1034, 2013.

Index

Printed in the United States
By Bookmasters